# Risks and Challenges of Hazardous Waste Management: Reviews and Case Studies

### Edited by

## Gabriella Marfe

*Department of Scienze e Tecnologie Ambientali Biologiche e Farmaceutiche,*
*University of Campania "Luigi Vanvitelli",*
*via Vivaldi 43,*
*Caserta 81100,*
*Italy*

### &

## Carla Di Stefano

*Department of Hematology,*
*"Tor Vergata" University,*
*Viale Oxford 81, 00133 Rome,*
*Italy*

# Graphical Abstract

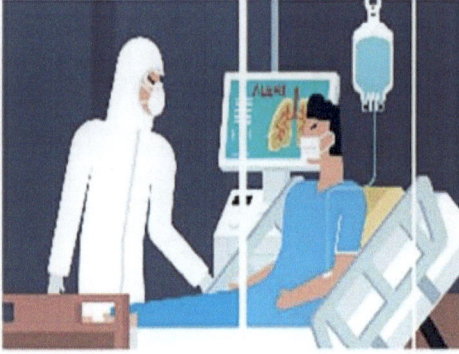

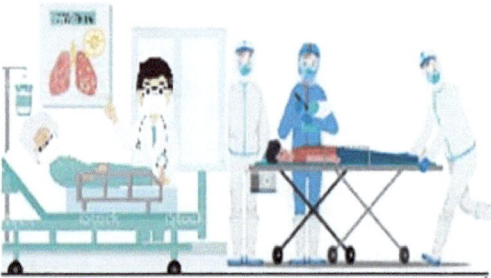

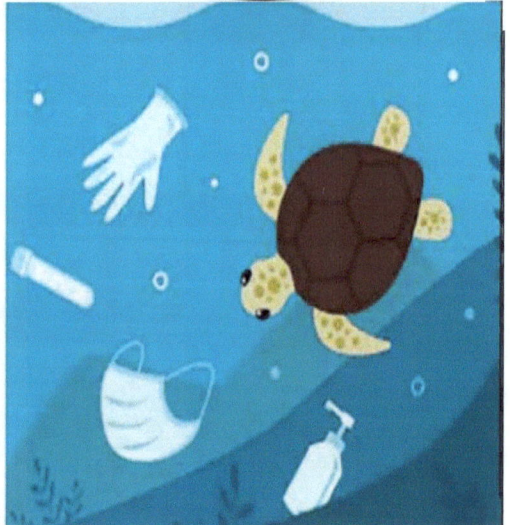

**Risks and Challenges of Hazardous Waste Management: Reviews and Case Studies**

Editors: Gabriella Marfe and Carla Di Stefano

ISBN (Online): 978-981-14-7246-6

ISBN (Print): 978-981-14-7248-0

ISBN (Paperback): 978-981-14-7247-3

need for a court order if at any point you breach any terms of this License Agreement. In no event will any delay or failure by Bentham Science Publishers in enforcing your compliance with this License Agreement constitute a waiver of any of its rights.

3. You acknowledge that you have read this License Agreement, and agree to be bound by its terms and conditions. To the extent that any other terms and conditions presented on any website of Bentham Science Publishers conflict with, or are inconsistent with, the terms and conditions set out in this License Agreement, you acknowledge that the terms and conditions set out in this License Agreement shall prevail.

**Bentham Science Publishers Pte. Ltd.**
80 Robinson Road #02-00
Singapore 068898
Singapore
Email: subscriptions@benthamscience.net

**BENTHAM SCIENCE**

# CONTENTS

# FOREWORD

Interest in solid and hazardous waste management is relatively recent, *i.e.*, in the last three decades, and is driven by regulations in most countries. It began with industrial hazardous waste, followed by municipal solid waste, and subsequently by many other categories of waste. This book features chapters discussing the implications of healthcare waste management and their impact on groundwater and other parts of the environment, as well as principles of as sustainable management and its application in the reuse and recycling of such kind of waste. Moreover, it includes examples of waste to energy. It also covers topics such as life cycle assessment as a tool for developing healthcare integrated waste management systems and an overview of waste management rules, illustrating the importance of technological inputs in the development of regulatory frameworks. This subject matter should be examined with a global standpoint in order to bridge the knowledge gap. Different chapters analyze the management of healthcare waste around the globe and specifically in different countries where it is in a deplorable state and stands out as a major risk factor to the health of both healthcare workers and individuals/communities in and around the healthcare facilities. There is a need to institutionalize healthcare waste management as part of the overall management system of a particular healthcare facility. In addition, healthcare waste management needs to intervene at other higher levels *i.e.* district, regional and national levels such that these levels will be positioned to provide the necessary or required support to facility and community levels in all matters related to healthcare waste management. The significant support expected from the national, regional and district levels includes, among others, the development of the required policies, legal/regulatory framework and ensures adequate budgetary allocations to meet healthcare waste management input requirements.

In this scenario, it is important to point out that medical care plays a vital role in our life and health. Still, the waste generated from medical activities represents a real problem of living nature and human world. For this reason, the appropriate methods for waste neutralization, recycling and disposal should be identified such management [1]. In particular, all processes should ensure both proper hospital hygiene and safety of health care workers and communities. Among biomedical waste, some chemical and pharmaceutical products that can cause poisoning by absorption through the skin or mucous membranes, by inhalation or by ingestion. Furthermore, they could provoke lesions of skin, eye, and respiratory mucosa. For example, chemical waste an or pharmaceuticals (such as antibiotics or other drugs, heavy metals, disinfectants and antiseptics) removed by the drainage system may cause toxic effects on ecosystems. Furthermore, the elimination of biomedical waste should be associated with safe waste management to protect the human health and the environment. Incineration of medical waste can be very dangerous since the plastic (containing chlorine) derived from such waste generates dioxin. Once formed, dioxin (that is carcinogenic) is able to bind to organic particles. Therefore, it is transported by wind and deposited on land and in water. The half-life of such compound is 25 to 100 years. Furthermore, dioxin it is able to link to nuclear DNA. Its formation is associated to potential cancer promoter, weak-delete immune response and other harmful effects on human health (such endometriosis, birth defects, low testosterone levels) and environment. Also, exposure to low dioxin concentrations causes negative effect on human health. The introduction of a sustainable system of biomedical waste management might allow to store a significant amount of hazardous biomedical waste in safe manner, and it will be possible to recover, treat, neutralize and recycle in terms of environmental protection. Therefore, waste recycling could play a crucial step in the reduction of earth resources. Furthermore, efficient health care waste management can be associated with the control of different diseases (hospital acquired infections), and reduction of community exposure to resistant bacteria. In addition, it could lead to the decrease of sepsis, and hepatitis transmission from dirty needles or uncleaned medical items. In this scenario, a sustainable management system of biomedical waste it is necessary to avoid the harm human health effects.

Furthermore, the coronavirus disease 2019 (COVID-19) pandemic has led to a great increse of medical and domestic waste. In this context, the safe managing of this waste plays a crucial role to successfully containing the disease. Therefore, the current pandemic brings a new challenge for medical waste management in every country.

In this context, some chapters of this book underline the relationship between adverse health effects due to waste management practices, in particular of hazardous waste, that potentially represent a public health issue in many countries because of growing waste production, inadequate waste management practices, lack of appropriate legislation and control systems, as well as of growing illegal hazardous waste transboundary movements [2, 3].

In this frame, various chapters describe different aspects on the occurrence and severity of health effects related to illegal waste disposal in Campania, an Italian region. Since 1980, this region has been characterized waste mismanagement [4 - 7], which led to the deterioration of land, as well as ground and surface water, also impacting air quality. In recent times, some oncologists, pathologists and toxicologists have reported that the continuous exposure of Campania citizens to toxic contaminants produced by the illegal dumping of waste in the region could become a big health issue. Furthermore, in these years, many grassroots movements against waste mismanagement are born in Campania. For example, in 2012, three main groupings of associations: Fires Coordination Committees, Campania Citizens for an Alternative Waste Management Plan, and Commons Net have created a social coalition called Stop Biocide [8]. Today, the citizens fight to obtain:

1. a better management of urban waste.
2. the remediation of the contaminated sites.
3. the halting of illegal waste trafficking.
4. a systematic health screening of the Campania population who live close to illegal dumping of toxic waste.

The faces of the children, who died and continue to die from cancer, represent the icons of biocide movement and they are shown during several demonstrations in the region. Furthermore, the movement has sent to the President of the Italian Republic and Pope Francis a lot of postcard with their faces printed on them to ask for direct intervention [9, 10].

Moreover, an article entitled "Triangle of death" linked to waste crisis, was published in The Lancet Oncology [11], and the authors reported a possible correlation between hazardous waste and high incidence of cancer in Naples and Caserta provinces. In this regard, people living close to illegal waste sited located in different municipalities of Naples and Caserta reported some adverse health effects. Many descriptive studies reported an early mortality rate (0–14 years) and congenital malformations. Furthermore, the cause-specific mortality and morbidity rates in Campania are very different when compared with the Italian national average. Future studies should be carried out to better understand the correlation between waste-related exposures and health profile of the Naples and Caserta provinces by analysis of mortality, hospital discharge records, cancer incidence, congenital malformations in newborns.

I recommend this book since it offers a broad look on the interaction between hazardous waste management and human health impacts. Through the chapters, it examines the way we affect and disrupt our health and the health of ecosystem around us. Now I believe that we have got to work together for the future. If we change now, we have an opportunity to decide what kind of world our children and our grand children and their children will grow up in.

Green Economics Institute Economics and Social and Environmental Justice http://www.greeneconomicsinstitutetrust.org/

# REFERENCES

[1]  McDougall F, White P, Franke M, Hundle P. Integrated Solid Waste Management: A Life Cycle Inventory OxfordBlackwell Science Edition2001.
[http://dx.doi.org/10.1002/9780470999677]

[2]  Harjula H. Hazardous waste: recognition of the problem and response. Ann N Y Acad Sci 2006; 1076: 462-77.
[http://dx.doi.org/10.1196/annals.1371.062] [PMID: 17119225]

[3]  Senior K, Mazza A. Triangle of death linked to waste crisis. Lancet Oncol 2004; S5(9):525-7. HYPERLINK. https://www.ncbi.nlm.nih.gov/pubmed/15384216

[4]  Piscitelli P, Santoriello A, Buonaguro FM, *et al.* Human Health Foundation Study Group. Incidence of breast cancer in Italy mastectomies and quadrantectomies performed between 2000 and 2005. J Exp Clin Cancer Res 2009; 19:28:86. https://www.ncbi.nlm.nih.gov/pubmed/19545369
[http://dx.doi.org/10.1186/1756-9966-28-86]

[5]  Altavista P, Belli S, Bianchi F, et al. Cause-specific mortality in an area of Campania with numerous waste disposal sites. Epidemiol Prev 2004; 28(6): 311-21.

[6]  Greyl L, Vegni S, Natalicchio M, Cure S, Ferretti J. The Waste Crisis in Campania, Italy 2010 Sito web CEECEC (disponibile in linea: http://www.ceecec.net/case-studies/waste-crisis-incampania-italy/

[7]  Trinca S, Comba P, Felli A, Forte T, Musmeci L, Piccardi A. Childhood mortality in an area of southern Italy with numerous dumping grounds: Application of GIS and preliminary findings. In Proceeding of the First European Conference "Geographic Information Sciences in Public Health", Sheffield, UK, 19–20 September 2001; p. 19.

[8]  De Rosa PS. The remaking of toxic territories: grassroots strategies for the re-appropriation of knowledge and space in the socio-environmental conflicts of Campania, Italy. Presented at the Political Studies Association Conference; Voice and Space: new possibilities for democracy in Southern Europe? Manchester, 2014, April 14-16.

[9]  Iengo I, Armiero M. The politicization of ill bodies in Campania, Italy. J Polit Ecol 2017; 24: 44-58.
[http://dx.doi.org/10.2458/v24i1.20781]

[10]  D'Alisa G, Germani AR, Falcone PM, Morone P. Political ecology of health in the land of fires: A hotspot of environmental crimes in the south of Italy. J Politic Ecol 2017; 24:59-86.

[11]  Senior K, Mazza A. Italian "triangle of death" linked to waste crisis. Lancet Oncol 2004; 5(9): 525-7.
[http://dx.doi.org/10.1016/S1470-2045(04)01561-X] [PMID: 15384216]

**Miriam Kennet**
Chartered Institute of Purchasing and Supply- MCIPS,
Alumna of the Month South Bank University,
London,
Editor Green Academic Journal,
Director CEO The Green Economics Institute
Head of United Nations Delegation to the COP Climate Conferences

# PREFACE

This book is written with the objective of providing all the essential information which are of utmost importance to hazardous waste management. The issues of environment protection have already spread far and wide and proper management of hazardous waste is one valuable contribution towards this global concern. The first chapters attempt to simplify the subject and to inculcate the valid concept of biomedical waste management effects.

From chapter one to three, the problems of healthcare waste management are discussed. Moreover, other authors illustrate some case studies of hazardous waste mismanagement that has caused a massive environmental damage in chapter four and five. Then, chapter six analyzes the emerging environmental and public health impacts of e-waste.

Finally, the last chapters describe the waste crisis in Campania. In this regard, the authors provide useful insight into single aspects of the waste system in Campania and their influences on human health impact.

**Gabriella Marfe**
Department of Scienze e Tecnologie Ambientali Biologiche e Farmaceutiche,
University of Campania "Luigi Vanvitelli",
*via* Vivaldi 43,
Caserta 81100,
Italy

**&**

**Carla Di Stefano**
Department of Hematology,
"Tor Vergata" University,
Viale Oxford 81, 00133 Rome,
Italy

# List of Contributors

| | |
|---|---|
| **Abdulkadir Aliyu** | Department of Urban and Regional Planning, Abubakar Tafawa Balewa University, Bauchi, Nigeria |
| **Annamaria Martuscelli** | Environmental Humanities Laboratory, Division of History of Science, Technology and Environment KTH Royal Institute of Technology, Stockholm, Sweden |
| **Atilio Savino** | Asociación para el Estudio de los Residuos Sólidos, Buenos Aires, Argentina |
| **Arvind Kumar Shukla** | School of Biotechnology and Bioinformatics, D.Y. Patil University, Navi Mumbai, 400614, Maharashtra, India<br>School of Biomedical Convergence Engineering, Pusan National University, Yangsan 50612, Korea<br>Inventra Medclin Biomedical Healthcare and Research Center, Katemanivli, Kalyan, Thane,421306, Maharashtra, India |
| **B.A Gana** | Department of Environmental Management Technology, Abubakar Tafawa Balewa University, Bauchi, Nigeria |
| **BenBouza Amina** | Natural Risks and Territory Planning Laboratory (LRNAT), Institute of Industrial Hygiene and Safety, Batna 2 University, Batna, Algeria |
| **Caputo Gaetano** | I.C. "F. Palizzi", Piazza Dante, 80026 Casoria Naples, Italy |
| **Carla Di Stefano** | Department of Hematology, "Tor Vergata" University, Viale Oxford 81, 00133 Rome, Italy |
| **Ernesto de Titto** | Universidad ISALUD, Buenos Aires, Argentina |
| **Gabriella Marfe** | Department of Scienze e Tecnologie Ambientali, Biologiche e Farmaceutiche, University of Campania "Luigi Vanvitelli,", Vivaldi 43, Caserta t81100, Italy |
| **Giulio Tarro** | Primario emerito dell'Azienda Ospedaliera "D. Cotugno", Napoli, Italy<br>University Thomas More U.P.T.M, Rome, Italy |
| **Houfani Roufaida** | Natural Risks and Territory Planning Laboratory (LRNAT), Institute of Industrial Hygiene and Safety, Batna 2 University, Batna, Algeria |
| **Lucio Righetti** | Environmental Humanities Laboratory, Division of History of Science, Technology and Environment KTH Royal Institute of Technology, Stockholm, Sweden |
| **Maryam Musa** | Department of Environmental Management Technology, Abubakar Tafawa Balewa University, Bauchi, Nigeria |
| **Salvatore Paolo De Rosa** | Environmental Humanities Laboratory, Division of History of Science, Technology and Environment KTH Royal Institute of Technology, Stockholm, Sweden |
| **S.V. A.R. Sastry** | Department of Chemical Engineering, MVGR College of Engineering (A), Vizianagaram, 535 005, India |
| **Sandhya Shukla** | Inventra Medclin Biomedical Healthcare and Research Center, Katemanivli, Kalyan, Thane, 421306, Maharashtra, India |

**Sefouhi Linda**      Natural Risks and Territory Planning Laboratory (LRNAT), Institute of Industrial Hygiene and Safety, Batna 2 University, Batna, Algeria

**Y.Y. Babanyara**    Department of Urban and Regional Planning, Abubakar Tafawa Balewa University, Bauchi, Nigeria

# Healthcare Waste: A Challenge for Best Management Practices in Developing Countries

**Ernesto de Titto**[1] and **Atilio Savino**[2,*]

[1] *Universidad ISALUD, Buenos Aires, Argentina*

[2] *Asociación para el Estudio de los Residuos Sólidos, Buenos Aires, Argentina*

**Abstract:** Healthcare waste (HCW) is the waste generated by the activities of healthcare facilities, educational institutions and medical research which is harmful to both human and animal health. About 10 to 15% of HCW presents hazardous characteristics, including a broad range of materials from sharps, used needles and syringes to soiled-dressings, body fluids or wastes contaminated by chemical and/or containing a high concentration of microorganisms. Such kind of waste requires very specific treatment to ensure proper final disposal. Its generation depends on different factors such as the economic development of the country and the type of service provided by the above-mentioned institutions. In this context, HCW management (HCWM) is a public health and environmental concern worldwide, especially for non-developed countries. Furthermore, HCWM is a complex and challenging process that covers a wide variety of actions, including segregation, minimization, previous treatment, packaging, temporary storage, collection, internal transportation and external storage of HCW. The first priority in this waste management should be the segregation and reduction in order to decrease the contaminated solid waste and to ensure selective collection. Furthermore, a great part of HCW can be recycled. In order to encourage successful best management practices, the results of a GEF-funded national development report headed by the Ministry of Health of Argentina are hereby exposed including proposed actions for training, guidelines, supervision, appropriate utility supply, management support and specific regulations to face future challenges. Improvements in the management system through HCW indicators may prove failures in segregation procedures, showing an opportunity for continual advances. To reduce potential problems that expose the healthcare facility staff, patients and their attendants to the risk of serious health hazards, there should be sufficient resource allocation, periodic training and strict supervision by stakeholders. Institutional planning for an efficient HCWM will assure HCF to both save money and provide a safe environment for patients and healthcare personnel.

**Keywords:** Best management practices, Healthcare facility, Healthcare waste, Healthcare waste management.

---

* **Corresponding author Atilio Savino:** Asociación para el Estudio de los Residuos Sólidos, Buenos Aires, Argentina; E-mail: asavino@ars.org.ar

**Gabriella Marfe & Carla Di Stefano (Eds.)**

## INTRODUCTION

As far as the WHO is concerned, let's remember that "by trying to achieve their goals of reducing health problems and eliminating potential risks to people's health, health services inevitably produce waste that can be dangerous on their own for health". Waste produced in the course of health-care activities has a greater potential for infection and injury than any other type of waste. Wherever waste is generated, safe and reliable methods for handling are therefore essential. Inadequate and inappropriate management of healthcare waste (HCW) can have serious public health consequences and a significant impact on the environment. Therefore, the proper management of HCW is a crucial component of environmental health protection" [1]. The rising demand for healthcare services in developing countries, at the world level, is causing a significantly high amount of HCW generation that requires both efficient management and proper disposal [2]. Special concern deserves limitations of healthcare facilities (HCF) –global denomination for places that provide healthcare, including hospitals, clinics, outpatient care centers, and specialized care centers, such as birthing centers and psychiatric care centers- to adequately segregate infectious or hazardous waste from ordinary domestic waste to treat this type of waste with proper technologies [3]. Healthcare waste management (HCWM) poses technical problems and is largely influenced by cultural, social and economic circumstances [1]. Developing countries need well-designed HCWM policies as well as a legislative framework and plans to achieve local implementation. Actions involved in the implementation of effective HCWM programs require multisectoral cooperation at all levels. The change has to be gradual and must be technically and financially sustainable in the long term.

Improving HCWM by enforcing knowledge and technical capacity for implementing and sustaining pollution-prevention measures, waste minimization and segregation practices are viable alternatives to the "business as usual" scenario. Developing countries need to adopt new strategies and treatment technologies that are affordable, that can be developed and serviced locally, requiring low-cost energy inputs, and are appropriate to HCF in urban and rural areas including, at worst, the need to operate at locations that may lack reliable electricity service and other utilities.

We will not discuss radioactive wastes since they are usually under strict rules and supervision established by the National Atomic Energy Organizations, aside from health authorities.

The remaining wastes can be grouped into two broad categories: bio-hazardous and chemicals, as summarized in Table **1**, which represent different types of risks as well as require different preventive actions.

Table 1. Hazardous healthcare waste categories.

| Hazardous Healthcare Waste Categories | | Descriptions and Examples |
|---|---|---|
| Bio - hazardous waste | Infectious | Waste suspected to contain pathogens and that poses a risk of disease transmission (*e.g.* waste contaminated with blood and other body fluids; laboratory cultures and microbiological stocks; waste including excreta and other materials that have been in contact with patients infected with highly infectious diseases in isolation wards) |
| | Pathological | Human tissues, organs or fluids; body parts; fetuses; unused blood products |
| | Sharps | Used or unused sharps (*e.g.* hypodermic, intravenous or other needles; auto-disable syringes; syringes with attached needles; infusion sets; scalpels; pipettes; knives; blades; broken glass) |
| Chemical waste | Chemical | Waste containing chemical substances (*e.g.* laboratory reagents; film developer; disinfectants that are expired or no longer needed; solvents; waste with high content of heavy metals, *e.g.* batteries; broken thermometers and blood-pressure gauges) |
| | Cytotoxic | Cytotoxic waste containing substances with genotoxic properties (*e.g.* waste containing cytostatic drugs; genotoxic chemicals) |
| | Pharmaceutical | Pharmaceuticals that are expired or no longer needed; items contaminated by or containing pharmaceuticals |

Source: Modified from WHO, 2015.

## THE HEALTH-CARE CONTEXT

A WHO assessment conducted in 22 developing countries in 2002 showed that the proportion of HCF that do not use proper waste disposal methods ranged from 18% to 64% [4].

Probably for its more evident and immediate impact, greater attention has been given to bio-hazardous waste. Even in facilities properly managing their waste, healthcare workers are exposed through a mucosal cutaneous or percutaneous route to accidental contact with human blood and other potentially infectious biological materials while carrying out their occupational duties.

A significant portion of the infections arising from blood-borne pathogens may be due to injuries from contaminated sharp objects (needles, blades, *etc.*) injuries. Literature has reported that incidence rates of sharps injuries range from 1.4 to 9.5 per 100 healthcare workers, resulting in a weighted mean of 3.7/100 healthcare workers per year. Sharps injuries have been reported to be associated with infective disease transmissions from patients to healthcare workers resulting in

0.42 hepatitis B virus (HBV) infections, 0.05-1.30 hepatitis C virus (HCV) infections and 0.04-0.32 Human Immunodeficiency Virus (HIV) infections per 100 sharps injuries per year [5]. The greatest risk is for nurses and auxiliary staff at the facility level.

Additional personnel at risk include landfill workers, waste pickers, scavengers and recyclers after HCW leaves the facility. Therefore, one key element required to install best practices for HCWM is to address the problem of the spread of blood-borne pathogens associated with improper handling and disposal of HCW. It has been long recognized by the WHO's policy on safe HCWM which calls for a long-term strategy "for the final disposal of HCW to prevent the disease burden" [1].

Furthermore, some years ago WHO estimated that overuse and unsafe use of healthcare injections caused annually 32% of new infections with HBV (21 million cases), 40% of new infections with HCV (2 million cases), and 5% of new infections with HIV (260,000 cases) [4, 6]. However, a significant portion of these could be linked to injections administered with devices reused in the absence of sterilization -practice that fortunately decreased in developing countries from 39.8 to 5.5% between 2000 and 2010 [7] - rather than to HCWM failures.

On the other hand, "emerging" diseases related to environmental exposure to chemicals have increased in recent years around the world [8]. Cancer and reproduction disorders (infertility, malformations, reproductive diseases), hormonal dysfunctions (diabetes, thyroid problems), immune diseases (dermatitis, allergies) and neurological diseases (learning problems, autism, hyperactivity, Alzheimer's, Parkinson's), among others, are increasingly linked to exposure to toxic chemicals. Environmental pollution to air, water and soil, as the result of the incorporation of toxic substances, wastes or residues incorrectly managed at the HCF, not only affects health workers but also far away populations. In fact, several studies on pharmaceutical contamination and HCW carried out have demonstrated that drugs represent a new class of contaminants including antibiotics, hormones, painkillers, tranquilizers, and chemotherapy products that are applied to cancer patients and can be found both on the surface and in deep waters, and also in the treatment plants of many hospitals [9]. International consensus on improving the management of chemical substances, including the health sector, has already been achieved through several specific conventions (Basel, Rotterdam, and Stockholm) now improved with the Strategic Approach to International Chemicals Management [10].

Among chemicals, a particular reference should be given to Mercury, a non-essential metal that has been frequently employed in health services in several devices and even as a component of pharmaceutical products, which does not fulfill any biochemical or nutritional function. In all its forms (elemental, organic and inorganic) it is an important environmental toxic and causes adverse effects on human health, especially in the fetal and infantile stages that are especially vulnerable to its harmful effects, highlighting toxicity neurological, renal and to the immune system.

Chemical waste generated at HCF, which may include the liquid waste from cleaning materials and disinfectants, expired and unused pharmaceutical products and cytotoxics are all considered hazardous waste products and they must be disposed of *via* an authorized system at approved sites (*e.g.*, industrial landfills). Thus, the main issue is the proper indoor management of chemical waste.

Dentistry waste deserves particular consideration since dentistry is commonly a private practice outside institutionalized HCF. Several studies indicate that the knowledge, attitude and practice of dental practitioners towards the management of dental waste still require strong improvement. Evaluations in Bangkok [11], New Zealand [12] and Kenya [13], for example, indicated that few dentists complied with all recommendations for the disposal of wastes with most waste being disposed of as domestic garbage.

Dental wastes, materials that have been utilized in dental clinics and are no longer accepted for use and are therefore discarded, may include biohazardous wastes, which may contain pathogenic organisms causing transmission of diseases such as HBV and HIV, especially in the presence of open wounds, as well as hazardous wastes, such as barium, cadmium, chromium, lead, polystyrenes, strontium, all of which may cause harm if improperly managed and disposed of.

## BEST MANAGEMENT PRACTICES

Many health professionals have only limited awareness about environmental health issues, including risks linked to toxic contaminants released into the environment. Moreover, waste management or the impacts of waste treatment choices are scarce or mostly absent in curricula in academic training programs for physicians, nurses, health specialists and administrators.

However, healthcare professionals are generally very receptive to environmental risk information and the extent of the harm they can cause. When made aware of this environmental health threat, they can be expected to support alternative waste management approaches that avoid generating and/or releasing toxic pollutants to the environment, as long as these alternatives are practical and do not compromise patient safety or care. Hence, the health sector has to be seen as a valuable ally in awareness-raising and advocacy with regard to minimizing or eliminating releases of contaminants to the environment.

Requirements for sustainable HCWM include both practices as well as technology. Adverse environmental and public health impacts of HCWM can be traced to both improper practices and the use of non-environmentally sound technologies. Poor practices that lead to high rates of HCW generation in HCF may include incomplete or even lack of segregation, unsafe handling of waste, dumping of untreated waste, extensive use of disposable materials, inadequate procedures for clean-up and containment of spills, weak inventory controls of time-sensitive pharmaceuticals and reagents, and inappropriate classification of non-infectious waste as bio-hazardous waste. Years ago incineration appeared as a promising solution to deal with HCW but experiences in many developing countries failed to fulfill adequate standards since the incinerators of choice cause objectionable smoke and odors, break down frequently and are difficult to properly operate and maintain. Moreover, small-scale incinerators often operate at temperatures below 800 degrees Celsius, thus leading to the production of dioxins, furans or other toxic pollutants as emissions and/or in bottom/fly ash. Last but not least, the installation of incinerators discourages efforts at segregation, recycling and waste minimization. The solution, therefore, must address both the practices and technologies applied.

There is no doubt that proper treatment of hazardous HCW must be part of an HCWM system, which must start by institutionalizing best management practices at HCF in order to minimize the production of HCW. In doing so, attention must be paid to the concerns for health services providers regarding both the quality and the costs of healthcare services.

Adopting good HCWM practices include pollution prevention and waste minimization through correct classification and segregation, proper containerization and color-coding, safe handling and collection of waste, labeling and signage, and proper storage, transport and final disposal of waste. It is evident that pollution prevention and waste minimization must be taken as priorities.

Waste minimization requires environmentally sound practices: reduction at source, material substitution, as well as safe reuse, recycling and composting of waste whenever possible. Limitations on these practices have been documented all over the world [14], including in Brazil [15 - 17], Botswana [18], Ethiopia [19, 20], Australia [21], Iran [22, 23], Lebanon [24] and other Asian [25] and African [26] countries.

Hazardous HCW (bio-hazardous and chemical waste) typically comprise about 25% or less of the total waste generated by HCF [1, 4]. However, rigorous segregation, as well as pollution prevention measures, can reduce significantly the amount of waste requiring special treatment. This is achievable by changing HCF practices through the development and effective implementation of effective plans with clear definitions of roles and responsibilities which must include changes in administrative policies established as well as installing motivational programs to promote process changes and regular training at all levels of the facility [27, 28]. It is also essential to install monitoring, periodic evaluation, continuous program improvements and full consideration of occupational safety and personal protection.

Alternative technologies suitable for HCW treatment must be capable of achieving international standards on microbial inactivation, being easy to operate and maintain and be affordable enough to become acceptable by HCF. Possible low-cost designs for resource-limited areas include locally made, small- to medium-scale pressure containers using electricity, gas, solar or other local fuels, as well as small manual and electrical shredders. Other alternative technologies available include autoclaves or retorts, with or without shredders to reduce waste volume and render unrecognizable HCW; advanced steam systems such as rotating autoclaves, combined pressurized steam-internal shredding units, hydroclaves, microwave systems, and alkaline hydrolysis to decompose tissues, anatomical and animal wastes, and possibly chemotherapeutic waste. These technologies became well-established and have been in operation for years. A number of other alternative technologies, such as chemical disinfection systems using chlorine and emerging technologies such as irradiation and plasma pyrolysis raise occupational safety or environmental issues including dioxin formation.

A summary of Best Management Practices is presented in Table **2**. In particular, mercury waste management requires the development of a mercury reduction plan that considers critical opportunities for material substitution, training, spill response and recovery, personal protection, segregation, containment, long-term engineered storage and encapsulation or amalgamation.

Mercury-free technologies include digital, glass alcohol, galinstan and tympanic thermometers, as well as aneroid sphygmomanometers. Mercury-free substitutes that are now commercially available can replace mercury-containing medical preservatives, fixatives and reagents. Increasing the demand for mercury-free products will help to lower the cost of these devices and mercury-free formulations.

Many HCF have already successfully switched to mercury-free thermometers and sphygmomanometers. A number of governments representing low-, middle- and high-income countries have also instituted policies for phasing out such devices in favor of accurate and affordable alternatives. Almost one hundred countries all around the world made a commitment to protect human health from anthropogenic emissions and releases of mercury and mercury compounds by signing the Minamata Convention on Mercury agreed in Kumamoto, Japan, in October 2013. WHO and numerous Health Ministries around the world are actively supporting the implementation of the Convention, including actions taken within the health sector. The 67th World Health Assembly in resolution WHA67.11 further affirmed this commitment. A phase-out date of 2020 for the manufacture, import and export of mercury thermometers and sphygmomanometers was included in the Convention. In 2015, WHO published a thoughtful guidance, available on the internet, to provide advice to Health Ministries on the leading role they will need to play in this regard [29].

Mercury in dentistry deserves particular attention. Since the XIX century amalgam became the dental restorative choice material to fill cavities caused by tooth decay, due to its low cost, ease of application, strength, and durability. Dental amalgam is a liquid mercury and metal alloy mixture commonly made of mercury (50%), silver (~22–32%), tin (~14%), copper (~8%) and other trace metals. Although there is a strong agreement to replace mercury-based amalgams for much less harmful materials already available in the market and legal regulations pushing in that direction, like the July 2018 prohibition by the European Union of using amalgam for dental treatment of children under 15 years and of pregnant or breastfeeding women [30], we are still far from complete replacement. Especially in less developed countries.

**Table 2. Best management practices for healthcare waste.**

| Healthcare Waste Category | Best Management Practices |
|---|---|
| **Bio-hazardous waste:** Sharps, materials contaminated with blood and bodily fluids, pathological waste, cultures and stocks, *etc.* | Minimization, classification, containerization, segregation, collection, color-coding, labeling, safe handling, storage, on-site and/or off-site transport, on- or off-site alternative treatment, disposal, *etc.* |
| **Chemical waste:** Mercury, chemotherapeutic waste, laboratory solvents, expired drugs, cleaning and maintenance chemicals, *etc.* | Minimization, inventory control, environmentally preferable purchasing, material substitution, segregation, safe handling, storage, solvent recovery, transport, encapsulation, *etc.* For mercury: spill kits, containment, storage, *etc.* |

*(Table 2) cont.....*

| Healthcare Waste Category | | Best Management Practices |
|---|---|---|
| **Non hazardous waste** | **Recyclable waste:** Paper, cardboard, glass, plastics, aluminum, wood, *etc*. | Waste minimization, environmentally preferable purchasing, source reduction, sorting, segregation, storage, collection, materials recovery, recycling, reuse, composting or vermicultura, disposal, *etc*. |
| | **Compostable waste:** Kitchen waste, yard waste, other organic waste. | |
| | **Non-recyclable municipal waste:** Other general waste that is not easily recyclable. | |

Source: Author's elaboration.

## TRENDS IN THE USE OF MEDICAL WASTE INCINERATORS

The use of medical waste incinerators (MWIs) appeared as a practical solution to deal with hazardous waste produced in HCFs. MWIs could be produced at different shapes so as to serve HCFs with different sizes and complexities and could be located locally thus minimizing waste transportation. Initial success in many industrialized countries was later hampered for difficulties to deal with toxic emissions to the environment. Consequently, MWI facilities fell into decline and were to be gradually phased out in many industrialized countries. For example, the number of MWIs in the United States dropped from 6,200 in 1998 to 111 in 2004 [31 - 33]; while in Canada dropped from 219 in 1995 to 120 in 2000 until December 2003 when the province of Ontario phased out all of its 56 MWIs further dropping the number of incinerators nationwide to 64 [34]. Similar trend can be seen in Europe. In Czech Republic, after its accession to the EU in 2004, all operational incinerators had to meet EU standards considering waste incineration and air protection ($0,1$ ng I-TEQ/m$^3$). As a result of this some of them were upgraded and continued to function while others were closed. In 2006, there were 32 only hazardous WI, in which HCW could be treated as well; in Slovenia, according to the Slovenian regulation No. 1520 of 30/95 Collection of law issued by Ministry of Health, until 2003 all infectious waste had to be dealt with mobile steam disinfection equipment. The decontaminated waste was consequently disposed on landfills. Incineration was allowed only for other hazardous hospital waste categories such as anatomical parts, cytostatics, and pharmaceuticals. In November 2003 a new law, regulating HCW was adopted and a new waste catalogue was prepared, based on EU legislation [35]. Germany completely shut down by 2002 all 554 on-site MWIs existing in 1984 using now about a thousand autoclaves used on-site as well as in four central treatment facilities, leaving only one central hospital-waste incinerator and two mixed-waste incinerators in the country; Ireland which years ago incinerated about half of its HCW in approximately 150 incinerators today use non-incineration technologies for practically of all HCW; Poland have closed down in recent years about a hundred of 186 MWI facilities; while Portugal changed from treating all of its HCW in 40 incinerators in 1995 to steam-based units processing 87% of its waste with only one incinerator remaining by 2004 [36, 37].

In 2004, WHO commissioned a screening-level health risk assessment for exposure to dioxins and furans from small-scale MWIs. The study found that the expected practice with their use resulted in unacceptable cancer risks under medium usage (two hours per week) or higher. The report concluded that small-scale MWIs should be employed as a transitional means of disposal for HCW [38].

On the other hand, the above-described increase in the amount of waste generated by HCF in developing countries results in the need of disposing of increasing quantities of waste leading many countries to install disposal methods based on combustion or incineration of HCW to contain pressing concerns about the spread of diseases caused by exposure to HCW. A diversity of technologies, including open burning and combustion devices ranging from "drum incinerators" to locally-constructed incinerators with no controls are being employed. Concerns about the spread of infectious diseases pushed the incorporation of imported small size or mid-size incinerators that have minimal controls and inadequately controlled large incinerators for central facilities.

However, in many cases, these new or upgraded facilities may still generate and release unintentional POPs at levels considerably higher than would be permitted in most donor countries. Although the trend of transferring obsolete technologies is no longer an acceptable framework for developing countries, often exported incinerators have a limited or nonexistent market in their countries of origin because they cannot satisfy domestic regulations related to air pollution, including the release of unintentional POPs. Furthermore, the HCFs do not meet recommended operating practices due to the absence of expertise to maintain and service these MWIs. Rising concern over the fate of HCW, along with the lack of strong regulatory and enforcement mechanisms, increases the possibility that small incinerators with poor designs and inadequate pollution controls keep working.

All these concerns do not invalidate incineration as an adequate technology to deal with waste if properly managed, as described by the International Solid Waste Association [39] nor imply environmental and/or human risks [40].

## ORGANIZING HCW MANAGEMENT

Different strategies should be adopted in each establishment adapted to the local context and the particularities of the HCF [41]. The main steps for implementing HCWM can be based on the following actions: 1. Define the policy of the establishment: 2. Document the commitment of the authorities; 3. Diagnose the state of the situation; 4. Plan and schedule; 5. Define the characteristics and requirements associated with each stage of waste management; 6. Communicate; 7. Coordinate with related processes or activities; 8. Track the system; and 9. Develop management documents. This process is summarized in Fig. (**1**).

| STEP | HEALTHCARE WASTE MANAGEMENT ACTIONS |
|---|---|
| 1 | DEFINE THE POLICY OF THE ESTABLISHMENT |
| 2 | DOCUMENT THE COMMITMENT OF THE AUTHORITIES |
| 3 | DIAGNOSE THE STATE OF THE SITUATION |
| 4 | PLAN AND SCHEDULE |
| 5 | DEFINE THE CHARACTERISTICS AND REQUIREMENTS ASSOCIATED WITH EACH STAGE OF WASTE MANAGEMENT |
| | 5.1 DEFINE SEGREGATION STRATEGIES |
| | 5.2 FITNESS STORAGE |
| | 5.3 FITNESS INTERNAL TRANSPORT |
| 6 | COMMUNICATE |
| 7 | COORDINATE WITH RELATED PROCESSES OR ACTIVITIES |
| | 7.1 WORKER'S HEALTH |
| | 7.2 INFRASTRUCTURE AND MAINTENANCE |
| | 7.3 TRAINING |
| | 7.4 ACTION FOR CONTINGENCIES |
| 8 | TRACK THE SYSTEM |
| | 8.1 EVALUATION GUIDE |
| | 8.2 INDICATORS |
| 9 | DEVELOP MANAGEMENT DOCUMENTS |
| | 9.1 INTERNAL MANAGEMENT MANUAL OF RESIDUES IN HEALTH CARE FACILITIES |
| | 9.2 REGISTRATIONS |
| | 9.3 PROCEDURES AND INSTRUCTIONS |

Source: Modified from Brunstein et al., 2014.

**Fig. (1).** Healthcare waste management program.

## Define the Policy of the Establishment

The HCF should define its policy in relation to waste, based on the spirit rising from national and local laws. This policy should include the vision of comprehensive management of HCW, the role of the HCF as a generator, the need for participation of workers, the community and political bodies, with an emphasis on training and updating knowledge and communication from management. The policy should include the development of management feasibility and sustainability strategies and their strengthening.

## Document the Commitment of the Authorities

The decision of the HCF authorities to implement the defined policy must be documented and endorsed by the appropriate authorities. This commitment is manifested through: a) the appointment of responsible personnel; the level of complexity of the HCF will define whether responsibility for waste management is assumed by a committee formed for this purpose or a single responsible, b) resource allocation; for materials as well as personnel and the training of all seniorities, c) communication to all staff of the establishment of the importance of the subject and the need to participate and get involved in achieving successful management.

## Diagnose the State of Situation

When taking the decision to implement a management or modify the existing one, it is advisable to establish a baseline that allows to know the status of the HCWM, its levels, classes and streams of waste generation for each of the areas of the establishment, through a standardized survey that will also serve as a reference for the measurement of improvements.

## Plan and Program

The realization of the policy adopted requires the formulation of a management plan that starts from the initial situation of the HCF, which is to be sustainable over time, with a clear definition of the purpose of the plan, its objectives and goals to be achieved. The management plan should also include its scope and duration, the resources to be allocated, the methods of monitoring, evaluation, control and review, the identification of responsible(s) for the implementation as well as references to the documents to be developed. Planning should be reviewed at regular time intervals, thus allowing its adequacy based on the degrees of knowledge and experience progress, and accompanied by the development of a programme that more accurately describes the actions to be taken, taking into account organizational and technical aspects, and the situation of human, physical and economic resources.

## DEFINE THE FEATURES AND REQUIREMENTS ASSOCIATED TO EACH OF THE RESIDUE MANAGEMENT STAGES

### Define Segregation Strategies

Segregation, understood as the appropriate selection of waste at the point of generation, according to the type of belonging and the characterization adopted, must be carried out by the person who generates the waste. This practice allows classifying waste by its danger, as well as sorting the materials that can be recycled and prevent it from contaminating when coming into contact with hazardous waste, unnecessarily increasing the volume of waste to be treated.

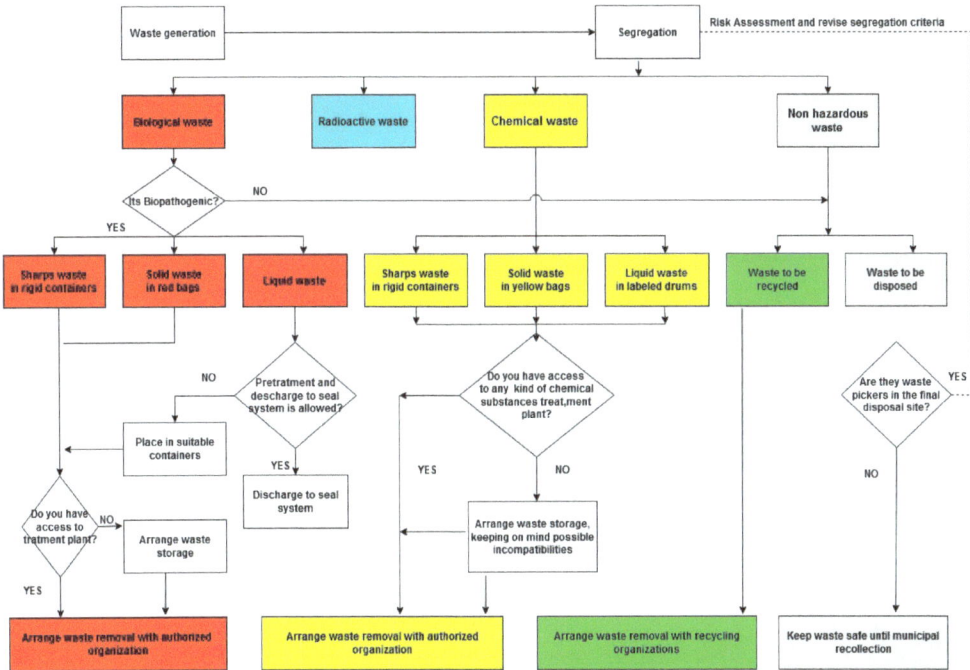

**Fig. (2).** Healthcare waste flux.

## Fitness Storage

Once the waste is generated and segregated, it will be temporarily arranged at different points of the establishment that must be located in properly signposted, ventilated and protected areas for direct rays from the sun or heat sources that receive the storage area name. Within the establishment, they can be classified into primary, intermediate and final storage. The location of the waste containment vessels, primary or intermediate, should be agreed between the staff and the Waste Committee or the Referrer, as appropriate. Once the locations of the containers have been agreed, a plan shall be drawn up indicating their location. The final storage within the establishment is the last place of collection of waste, until its internal treatment by the institution or withdrawal by the external operator that will perform the treatment and final disposal. The final storage site should be exclusively for waste and have, if possible, different premises for each type of waste. Final storages must have enough space to make transport, cargo and heavy duty. The premises identified and signposted outside must remain closed, with restricted access to personnel not related to HCWM. Its location on the property must allow direct access by external operators who remove the waste. In addition, the route of waste from inside the premises to its cargo in the collection trucks should be minimized, avoiding contact with staff not involved in the task and the general public. Last but not least, location should not affect the biosecurity, scenic quality, hygiene and safety of other sectors of the establishment and its environment.

## Fitness Internal Transport

The objective of internal collection and transport is to reduce the time waste spends at generation points in such a way as to minimize the risks of exposure to waste by workers, patients and the public. A transport logistics should be established, which involves determining the frequency and route of the transport to minimize the passage of carriages and bags by patient assistance areas and/or other clean or restricted locations. To establish the collection frequency, the generation levels by service or area and the requirements of the generation levels must be analyzed. Once the waste is segregated the collection should be differentiated, thus avoiding the mixing of the residual currents. If there is no possibility of differentiated dirty-clean traffic, travel schedules should be established, avoiding transit through the areas and times of higher density of people (for example, some restrictive schedules will be medical tours, office schedules, patient transfer, visits, inpatient meal service). In case there is no exclusive elevator for general services, schedules of use for the transfer of waste should be established, posters being placed informing the exclusive use at that time and performing the subsequent hygiene of the elevator. Once the locations of the storage spaces and transport routes have been agreed upon, a plan is to be drawn up to be exposed on the walls. For those HCFs that have intermediate storage, the collection will be divided into two: primary collection from primary to intermediate storage, and secondary collection from intermediate storages to the depot final storage.

## COMMUNICATE

The communication of the contents and requirements of HCWM arises as a need to motivate the participation and commitment of the workers and users of the establishments, thus contributing to implementing the changes to achieve the objectives of management. The uncertainties and stresses due to the implementation of new practices associated with HCWM are remedied by adequate communication to workers. Internal communication must be bidirectional, between all workers and the levels responsible for HCWM. Workers should not be mere recipient subjects, channels should be generated to allow communication. Communication can be developed verbally, visually and through written documents, depending on the possibilities of each establishment. The external communication of the HCF with the community is essential to inform about the practices and real risks of the waste generated in the establishment.

## COORDINATE WITH RELATED PROCESSES OR ACTIVITIES

The proper implementation, maintenance, monitoring, control and evaluation of HCWM activities requires the existence of other processes or activities and other areas that support and accompany their development, for example, training, selection of personnel, purchases, computer science, worker health, environment, external third parties (enabled operators), maintenance or infrastructure, patient safety, biosecurity and contingency action, among others.

### Worker's Health

Health and safety in HCWM, due to its potentially harmful consequences not only for workers but also for the general population and the environment, should be one of the main concerns in HCWM programs. It, therefore, requires actions between the Waste Committee and the health of the worker section, which include risks related to tasks at the various stages of management, and their prevention during waste work, taking into account that the target population within the HCF includes health workers, outpatients and hospitalized patients and others concurrent to the establishment; damage to health, both by occupational accidents and occupational diseases, with the corresponding Medical Surveillance; skills and training of personnel for waste-related tasks; communication of risks, ways of prevention and protection, the correct way of performing tasks to all workers and the selection of personal protection elements.

### Infrastructure and Maintenance

The prevention of damage associated with the presence of biopathogenic or chemical waste, implies the adoption of physical barriers, correct infrastructure conditions, adequate quality of construction and materials used. The necessary space requirements for all stages of waste management should be considered as well.

### Training

Generally, decisions made on a daily basis in the organization of tasks are usually based on the routine, historical modality of the establishment, or the know-how of the actors involved. To generate change, action must be taken on factors such as organizational climate and dynamics, staff training and motivation for improvement in waste management. The waste management officer should develop a training program, aimed at promoting healthy practices for the worker and being environmentally friendly, implementing it and ensuring their support. The minimum content should cover legal aspects (local and national); duties and workers' rights; knowledge and prevention of risk factors related to waste handling (chemical, biological, ergonomic, *etc.*); procedures for each waste activity; work instructions and safety sheets; features and use of personal protective elements; contingencies and occupational diseases. All the staff of the establishment must be trained, with the nuances relevant to the responsibilities that fit each activity. For those workers directly related to waste management, specific and continuous training should be carried out. New staff entering the facility or those who change work areas must be trained regardless of the annual training plan. Short-term training for waste generators should be carried out in order to review segregation practices in the everyday environment. Awareness-raising of the establishment's members, coupled with the acquisition of specific knowledge on adequate waste management, will in turn lead to improved protection of workers and in biosecurity conditions, to more environmental care standards and the achievement of cost optimization.

## Action for Contingencies

HCF staff in general and especially HCWM personnel should be trained to deal with management contingencies. The overall contingency plan for HCWM should integrate the facility's emergency plan. Its objective is to determine the guidelines and actions envisaged to increase the capacity to respond to any contingency, whether natural or man-made, minimizing the consequences that could be derived from them. Established protocols that define the roles, missions and functions of the staff are necessary at the facility for action in particular identified situations as possible to occur. There should also be a strategic and operational structure to control contingencies and minimize their negative consequences. This contingency plan should include protocols that face action if accidents occur, spills and leaks, cleaning and disinfection methods and/or decontamination, chain of notification of the fact, guidelines for declaring the beginning and end of the contingency, resources necessary for the action, first aid and replenishment of resources used. Actions must comply with the legal requirements of HCWM, minimize the exposure of workers during the intervention, and minimize the impact on the establishment, patients, other people outside the fact and the environment. Following the relevant intervention in the event of a contingency, a detailed summary of the facts to include should be made: adopted procedures, need for environmental and health follow-ups, recommendations for improvements, research causes of contingency and other comments and those deemed relevant.

## TRACKING THE SYSTEM

HCWM should be systematically monitored to assess progress in implementing best practices, detect the potential for improvement and identify those corrective or preventive actions that need to be implemented. The evaluation of the system allows to observe what happens at the various stages of management, evaluate its results, determine the degree of compliance with objectives, highlight economic and administrative efficiency, correct deviations and establish warning and intervention guidelines for diversions to implement corrective or preventive measures.

### Evaluation Guide

The tool used to make the diagnosis of HCWM, applied periodically, allows tracking the activities implemented and evaluating their effectiveness or needing improvement.

### Indicators

The use of specific indicators for HCWM allows the monitoring and control of activities and provides information on certain parameters that make the internal HCWM process and allows evaluating the results and impacts achieved. The Table **3** presents some HCWM indicators, selected based on the observation of repeated situations, which highlight critical points in management that can have health, legal or economic consequences.

**Table 3. Healthcare waste management indicators.**

| STAGE | INDICATOR | DESCRIPTION | EQUATION | SOURCE |
|---|---|---|---|---|
| STRUCTURE | Availability of supplies required for HCWM | Availability of supplies required for HCWM in amount, quality and opportunity | $\dfrac{\text{N° of supplies received as required}}{\text{N° of supplies required}}$ | Supplies receiving form |
| | Training | Measures the fraction of personnel receiving training for HCWM | $\dfrac{\text{Facility workers receiving training in HCWM (N°)}}{\text{Facility workers (N°)}}$ | HCWM training records Facility personnel |

*(Table 3) cont.....*

| STAGE | INDICATOR | DESCRIPTION | EQUATION | SOURCE |
|-------|-----------|-------------|----------|--------|
| RESULTS | Waste generation according to separation criteria | Measures the waste fractions distribution according to separation criteria established | $\dfrac{\text{Waste per type (Kg)}}{\text{Total amount of waste generated (Kg)}}$ | Non hazardous waste weighting form Hazardous waste weighting form Chemical waste weighting form Recyclable waste weighting form |
| | Non hazardous waste generated | Measures the amount of nonhazardous waste generated in a facility daily | $\dfrac{\text{Hazardous waste (Kg)}}{\text{Occupied bed/day}}$ $\dfrac{\text{Hazardous waste (Kg)}}{\text{Patient or practice/day}}$ | Non hazardous waste weighting form Facility statistics |
| | Chemical waste generated | Measures de amount of chemical waste generated in a facility daily | $\dfrac{\text{chemical waste (Kg)}}{\text{Occupied bed/day}}$ $\dfrac{\text{chemical waste (Kg)}}{\text{Patient or practice/day}}$ | Chemical waste weighting form Facility statistics |
| | Recyclable waste generated | Measures de amount of recyclable waste generated in a facility daily | $\dfrac{\text{Recyclable waste (Kg)}}{\text{Total amount of non hazardous waste generated (Kg)}}$ | Non hazardous waste weighting form Recyclable waste weighting form |
| IMPACT | Incidence of sharp injuries in HCWM | Number of sharp injuries in exposed workers | $\dfrac{\text{Sharp injuries (N°) in HCW workers}}{\text{HCW workers (N°)}}$ | Injuries registry Facility personnel |
| | Incidence of sharp injuries per occupational profile | Number of sharp injuries in exposed workers separated according to his/her occupational category | $\dfrac{\text{Sharp injuries in exposed workers per occupational profile (N°)}}{\text{Total number of sharp injuries (N°)}}$ | Injuries registry Facility personnel |
| | Incidence of chemical incidents | Measures de number of incidents involving chemicals | $\dfrac{\text{Incidents with chemicals (N°)}}{\text{Facility workers}}$ | Emergency registry Facility personnel |

*(Table 3) cont.....*

| STAGE | INDICATOR | DESCRIPTION | EQUATION | SOURCE |
|---|---|---|---|---|
| SATISFACTION | Personnel perception | Evaluates personnel perception on HCWM | $\dfrac{\text{Satisfactory answers (N°)}}{\text{Inquiries performed (N°)}}$ | Satisfaction inquiries |

Source: Modified from Brunstein *et al.*, 2014.

## DEVELOP MANAGEMENT DOCUMENTS

Management is facilitated with the development of a series of documents that allow its implementation of monitoring and control, internal communication and satisfaction of legal requirements. Some of them, which are not found in pre-established formats in most laws, are:

### Internal Management Manual of Residues in Health Care Facilities

Each HCF must develop its own Internal Waste Management Manual attending the particularities of the establishment.

### Registrations

There is a variety of records and legal provisions that must be complied with and documented for proper waste management. However, many of which are not strictly defined by the rules, but are associated with them. These records and provisions are represented by the manifests of transport, treatment and final disposal of waste, waste treatment certificate, registration of delivery of personal protection elements, registration of training, certificates of qualifications, registrations as a generator, coverage for occupational accidents and diseases, hygiene and safety checks or working conditions, possession of carcinogenic substances, chemical precursors, or medicines, among others. Those records that are specific to HCWM require the definition of responsibility for its elaboration, implementation and storage, as well as the definition of the time and sites of saving.

### Procedures and Instructions

The procedures associated with HCWM should be standardized, with precise and thorough instructions associated with each stage, for example for the collection of waste, cleaning of transport cars, cleaning of areas of storage, the use of personal protective elements, contingency action, hand washing.

### CONCLUDING REMARKS

At present, HCWM is a public health and environmental concern worldwide, particularly in developing countries; omissions in waste management might produce potentially health and environmental negative impacts.

Mismanaging hazardous waste affects all the community, particularly healthcare providers. The general and hazardous waste types should be properly segregated at their source of generation, which represents a big challenge for health services around the world. The segregated HCW types are required to be collected separately using waste-collecting utilities designed for each type of HCW. All hazardous HCW types generated from the HCFs should be stored in utility rooms prepared for cleaning equipment, dirty linen and waste storage. Each HCF should formulate its HCWM plan and guideline document. A key role should be assigned to training, since it enhances awareness of knowledge and practices among HCW handlers to achieve the desired objectives. Training sessions should not become merely a one-time activity but should be a continuous process in every HCF.

In most developing countries, incineration is frequently used as an HCW treatment method for hazardous HCW before the final disposal. However, incinerators are often operated under sub-optimal conditions and mostly with untrained personnel. Thus, harmful substances can be released into the environment due to inadequate incineration.

HCWM is a complex and challenging process. Lack of training, accessible guidelines, regular supervision, appropriate utility supply, management support, and specific rules/regulations are major challenges for implementing effective waste management systems. Improvements in the management system through HCW indicators may highlight failures in segregation procedures, showing an opportunity for continuous advances. There should be sufficient resource allocation, periodic training and strict supervision by the stakeholders. Institutional development of HCWM plans will assure HCF to both save money and provide a safe environment for patients and healthcare personnel.

**NOTES**

[1] The content of this section was originally developed by L Brunstein, D Alfano, R Benítez, F Chesini, S Sagardoyburu; M Montecchia and E de Titto, from the Ministry of Health-Argentina, in the GEF-funded project DEMONSTRATING AND PROMOTING BEST TECHNIQUES AND PRACTICES FOR REDUCING HEALTH-CARE WASTE TO AVOID ENVIRONMENTAL RELEASES OF DIOXINS AND MERCURY developed by the Ministry of Health of Argentina with the partnership of PAHO-WHO, the technical support of the NGO Health Care without Harm and the administrative support of UNDP.

## CONSENT FOR PUBLICATION

Not applicable.

## CONFLICT OF INTEREST

The authors confirm that the content of this chapter has no conflict of interest.

## ACKNOWLEDGEMENTS

We acknowledge D. Alfano, R. Benítez, L. Brunstein, F. Chesini, M.F. Montecchia, and S. Sagardoyburu –from Ministry of Health, Argentina- and A. Iwanaga, J. Emmanuel -from UNOPS, USA- for their participation in the elaboration of HCW Management Guidelines and Gabriella Marfe for her help in language editing of the draft.

### ACRONYMS

| HBV | Hepatitis B Virus |
| --- | --- |
| HCV | Hepatitis C Virus |
| HCW | Health-care waste |
| HCWM | Health-care waste management |
| HIV/AIDS | Human Immunodeficiency Virus/Auto-Immune Deficiency Syndrome |
| MWI | Medical waste incinerator |
| PAHO | Pan American Health Organization |
| POPs | Persistent organic pollutants |
| UNDP | United Nations Development Programme |
| UNEP | United Nations Environment Programme |
| WHO | World Health Organization |

## REFERENCES

[1]    WHO. Safe management of wastes from health-care activities. Chartier Y, Emmanuel J, Pieper U, Eds. Geneva, Switzerland 2014.

[2]    UUNN. The Sustainable Development Goals Report 2018. https://unstats.un.org/sdgs/report/ 2018/overview/

[3]    de Titto E, Savino AA, Townend WK. Healthcare waste management: The current issues in developing countries. Waste Manag Res 2012; 30(6): 559-61.
[http://dx.doi.org/10.1177/0734242X12447999] [PMID: 22692903]

[4]    WHO. Safe health-care waste management 2004.https://www.who.int

[5]    Elseviers MM, Arias-Guillén M, Gorke A, Arens HJ. Sharps injuries amongst healthcare workers: review of incidence, transmissions and costs. J Ren Care 2014; 40(3): 150-6.
[http://dx.doi.org/10.1111/jorc.12050] [PMID: 24650088]

[6] Hauri AM, Armstrong GL, Hutin YJ. The global burden of disease attributable to contaminated injections given in health care settings. Int J STD AIDS 2004; 15(1): 7-16.
[http://dx.doi.org/10.1258/095646204322637182] [PMID: 14769164]

[7] Hayashi T, Hutin YJF, Bulterys M, Altaf A, Allegranzi B. Injection practices in 2011-2015: a review using data from the demographic and health surveys (DHS). BMC Health Serv Res 2019; 19(1): 600.
[http://dx.doi.org/10.1186/s12913-019-4366-9] [PMID: 31455315]

[8] Prüss-Üstün A, Corvalán C. Preventing disease through healthy environments Towards an estimate of the environmental burden of disease. Geneva, Switzerland: World Health Organization 2006.http://www.who.int/quantifying_ehimpacts/publications/preventingdisease5.pdf

[9] Quesada Peñate I, Jáuregui Haza UJ, Wihelm AM, Delmas H. Contaminación de las aguas con productos farmacéuticos 2009.http://www.redalyc.org/articulo.oa?id=181221662005

[10] SAICM. 2012.www.saicm.org

[11] Punchanuwat K, Drummond BK, Treasure ET. An investigation of the disposal of dental clinical waste in Bangkok. Int Dent J 1998; 48(4): 369-73.
[http://dx.doi.org/10.1111/j.1875-595X.1998.tb00698.x] [PMID: 9779120]

[12] Treasure ET, Treasure P. An investigation of the disposal of hazardous wastes from New Zealand dental practices. Commun Dent Oral Epidemiol 1997; 25(4): 328-31.
[http://dx.doi.org/10.1111/j.1600-0528.1997.tb00948.x] [PMID: 9332812]

[13] Osamong LA, Gathece LW, Kisumbi BK, Mutave RJ. Management of dental waste by practitioners in nairobi, Kenya. African J Oral Health 2005; 2(1 & 2): 24-9.

[14] Hossain MS, Santhanam A, Nik Norulaini NA, Omar AK. Clinical solid waste management practices and its impact on human health and environment--A review. Waste Manag 2011; 31(4): 754-66.
[http://dx.doi.org/10.1016/j.wasman.2010.11.008] [PMID: 21186116]

[15] Moreira AM, Günther WMR. Assessment of medical waste management at a primary health-care center in São Paulo, Brazil. Waste Manag 2013; 33(1): 162-7.
[http://dx.doi.org/10.1016/j.wasman.2012.09.018] [PMID: 23122204]

[16] Ream PS, Tipple AF, Salgado TA, *et al.* Hospital housekeepers: Victims of ineffective hospital waste management. Arch Environ Occup Health 2016; 71(5): 273-80.
[http://dx.doi.org/10.1080/19338244.2015.1089827] [PMID: 26359679]

[17] Delmonico DVG, Santos HHD, Pinheiro MAP, de Castro R, de Souza RM. Waste management barriers in developing country hospitals: Case study and AHP analysis. Waste Manag Res 2018; 36(1): 48-58.
[http://dx.doi.org/10.1177/0734242X17739972] [PMID: 29153036]

[18] Mbongwe B, Mmereki BT, Magashula A. Healthcare waste management: current practices in selected healthcare facilities, Botswana. Waste Manag 2008; 28(1): 226-33.
[http://dx.doi.org/10.1016/j.wasman.2006.12.019] [PMID: 17350817]

[19] Mesfin A, Worku W, Gizaw Z. Assessment of health care waste segregation practice and associated factors of health care workers in Gondar University Hospital, North West Ethiopia, 2013. Univ J Public Health 2014; 2(7): 201-7.
[http://dx.doi.org/10.13189/ujph.2014.020703]

[20] Yazie TD, Tebeje MG, Chufa KA. Healthcare waste management current status and potential challenges in Ethiopia: a systematic review. BMC Res Notes 2019; 12: 285.
[http://dx.doi.org/10.1186/s13104-019-4316-y]

[21] Peng Bi , Tully PJ, Boss K, Hiller JE. Sharps injury and body fluid exposure among health care workers in an Australian tertiary hospital. Asia Pac J Public Health 2008; 20(2): 139-47.
[http://dx.doi.org/10.1177/1010539507312235] [PMID: 19124307]

[22] Dehghani MH, Azam K, Changani F, Dehghani Fard E. Assessment of medical waste management in

educational hospitals of Tehran university medical sciences. Iran J Environ Health Sci Eng 2008; 5(2): 131-6.

[23]   Askarian M, Momeni M, Danaei M. The management of cytotoxic drug wastes in Shiraz, Iran: an overview of all government and private chemotherapy settings, and comparison with national and international guidelines. Waste Manag Res Epub 2013.
[http://dx.doi.org/10.1177/0734242X13476747]

[24]   Musharrafieh UM, Bizri AR, Nassar NT, *et al.* Health care workers' exposure to blood-borne pathogens in Lebanon. Occup Med (Lond) 2008; 58(2): 94-8.
[http://dx.doi.org/10.1093/occmed/kqm139] [PMID: 18211911]

[25]   Ananth AP, Prashanthini V, Visvanathan C. Healthcare waste management in Asia. Waste Manag 2010; 30(1): 154-61.
[http://dx.doi.org/10.1016/j.wasman.2009.07.018] [PMID: 19726174]

[26]   Emilia A, Julius N, Gabriel G. Solid medical waste management in Africa. Afr J Environ Sci Technol 2015; 9: 244-54.
[http://dx.doi.org/10.5897/AJEST2014.1851]

[27]   Chowdhary A. Study of knowledge, behavior and practice of biomedical waste among health personnel. Int J Community Med Public Health 2018; 5(8): 3330-4.
[http://dx.doi.org/10.18203/2394-6040.ijcmph20183056]

[28]   Hosny G, Samir S, El-Sharkawy R. An intervention significantly improve medical waste handling and management: A consequence of raising knowledge and practical skills of health care workers. Int J Health Sci (Qassim) 2018; 12(4): 56-66.
[PMID: 30022905]

[29]   WHO. Developing national strategies for phasing out mercury-containing thermometers and sphygmomanometers in health care, including in the context of the Minamata Convention on Mercury: key considerations and step-by-step guidance. Geneva, Switzerland 2015.

[30]   2017.http://data.europa.eu/eli/reg/2017/852/oj

[31]   Hospital Waste Combustion Study—Data Gathering Phase, US Environmental Protection Agency, December 1988.

[32]   HMIWI Facility and Emission Inventory—English Units (draft), US Environmental Protection Agency, January 1 2004.

[33]   Dioxin and Furan Inventories: National and Regional Emissions of PCDD/PCDF, United Nations Environment Programme, Geneva, Switzerland, May 1999; B. Sibbald, Canadian Medical Association Journal, 164(4), February 20, 2001; and E. Lopes and S. Rossi, Hospital News, February 2003.

[34]   Yearbook of the Czech Ministry of Environment for 2003, issued November 2004.

[35]   Petrova S, J Petrlík. Healthcare Waste Treatment: The Czech Republic and Slovenia in Comparative Perspective. 2008. Available at https://www.google.com/ url?sa=t& rct=j&q=&esrc=s&source=web& cd=&cad=rja&uact=8&ved=2ahUKEwiA_4mT_b7rAhUeHbkGHYMjBfYQFjAGegQICBAB&url=ht tps%3A%2F%2Fwww.researchgate.net%2Fprofile%2FJindrich_Petrlik2%2Fpublication%2F3241329 36_Healthcare_Waste_Treatment_The_Czech_Republic_and_Slovenia_in_Comparative_Perspective %2Flinks%2F5abfe4f6aca27222c759b682%2FHealthcare-Waste-Treatment-The-Czech-Rep-blic-and-Slovenia-in-Comparative-Perspective.pdf&usg=AOvVaw3UmoYIgo5XKYsEFc_NvVpf

[36]   Emmanuel J, Hrdinka C, Głuszyński P, *et al.* Non-Incineration Medical Waste Treatment Technologies in Europe. Health Care without Harm Europe (Ed.), 2004; 44pages. Available at https://www.google.com/url?sa=t&rct=j&q=&esrc=s& source=web&cd=&cad=rja&uact=8&ved=2ahUKEwj8weG9n8brAhWLEbkGHczuDIAQFjAAegQIB hAB&url=https%3A%2F%2Fwww.env-health.org%2FIMG%2Fpdf%2Faltech_Europe_updated_version_10_12_2004.pdf&usg=AOvVaw2Pg DKTGiBU6dWkKVfE18S2

[37]    Gluszynski P. Emmanuel J. Best Environmental Practices and Alternative Technologies for Medical Waste Management. In the Eighth International Waste Management Congress And Exhibition, Institute of Waste Management of Southern Africa-Botswana Chapter 25, Kasane, Botswana, June 2007. Available at https://www.google.com/url?sa=t&rct=j&q=&esrc=s&source=web&cd=& cad=rja&uact=8&ved=2ahUKEwju9aGKjcTrAhXwJrkGHfKUAyo4WhAWMAF6BAgDEAE&url=ht tps%3A%2F%2Fnoharm.org%2Fsites%2Fdefault%2Ffiles%2Flib%2Fdownloads%2Fwaste%2FMed Waste_Mgmt_Developing_World.pdf&usg=AOvVaw2kV0xcwqpDTaBzAZiEnscF

[38]    Batterman S. Assessment of small-scale incinerators for health care waste. Report prepared for the Protection of the Human Environment. Geneva-Switzerland: World Health Organization 2004. Available at http://www.who.int/water_sanitation_health/medicalwaste/en/smincinerators.pdf

[39]    ISWA. Waste to Energy in Low and Medium Income Countries, ISWA Guidelines 2013. Available at https://www.google.com/url?sa=t&rct=j&q=&esrc=s&source=web&cd=1&cad=rja&uact=8&ved=0ah UKEwi33-7LwvnbAhWGi5AKHT5KA30QFggsMAA       &url=https%3A%2F%2Fwww.iswa.org %25%202Findex.php%3FeID%3Dtx_bee4mecalendar_download%26eventUid%3D243%26filetype% 3Dpublic%26filenum%3D2&usg=AOvVaw3MUNvP7fcTX1T8KyMePJH  "https://www.google.com/ url?sa=t&rct=j&q=&esrc=s&source=web&cd=1&cad=rja&uact=8&ved=0ahUKEwi33- 7LwvnbAhWGi5AKHT5KA30QFggsMAA&url=https%3A%2F%2Fwww.iswa.org% 2Findex.php%3FeID%3Dtx_bee4mecalendar_download%26eventUid%3D243%26filetype%3Dpublic %26filenum%3D2&usg=AOvVaw3MUNvP7fcTX1T8KyMePJH

[40]    de Titto E, Savino A. Environmental and health risks related to waste incineration. Waste Manag Res 2019; 37(10): 976-986. [http://dx.doi.org/10.1177/0734242X19859700]

[41]    Brunstein L, Chesini F, Montecchia MF. *et al.* Herramientas para la gestión de residuos en establecimientos de atención de la salud (Tools for Health Care Waste Management), Serie Temas de Salud Ambiental N° 22, 2014, Ministry of Health-Argentina, 155 pp. ISBN 97B-950-38-0194-9. Available at http://bancos.salud.gob.ar/recurso/ herramientas -para-la-gestion-de-resid- os-en-establecimientos-de-atencion-de-la-salud

**CHAPTER 2**

# A Review of the Knowledge, Attitude, and Practices of Healthcare Wastes Workers (HCWS) on Medical Waste in Developing Countries

**Y.Y. Babanyara[1,\*], Abdulkadir Aliyu[1], B.A Gana[2] and Maryam Musa[2]**

[1] *Department of Urban and Regional Planning, Abubakar Tafawa Balewa University, Bauchi, Nigeria*

[2] *Department of Environmental Management Technology, Abubakar Tafawa Balewa University, Bauchi, Nigeria*

**Abstract:** Medical care activities can produce various types of risks (hazardous) wastes. Poor management of these wastes can lead to environmental pollution and health risks to healthcare personnel, patients, and the community at large. Adequate knowledge, attitudes, and practices of managing medical waste are vital. This paper reviews the main issues in medical waste management by healthcare workers in developing countries. Results from reviewed literature showed that in developing countries, Medical waste management is inefficient. Knowledge and awareness concerning safe medical waste management are inadequate as a result of lack of or absence of training for medical waste management personnel, absence of waste management and disposal systems, lack of safety equipment and immunization in most of the health centers. This paper concludes by recommending ways by which poor medical waste management can be ameliorated in healthcare centers.

**Keywords:** Attitudes, Awareness, Healthcare, Healthcare waste workers, Hospital, Knowledge of Medical waste.

## INTRODUCTION

Medical center or hospital is where people go irrespective of their race, age, gender, and faith, to find a cure for their illnesses [1, 2]. Healthcare service delivery in medical centers generates hazardous waste known as 'Medical waste (MW), Clinical (CW), Healthcare (HCW) or Biomedical (BMW) wastes' [3], these terms are often been used synonymously. Lack of awareness regarding the dangers or risks associated with the refuse generated by health centers on mankind and the environment abound in the general population [4].

\* **Corresponding author Y.Y Babanyara:** Department of Urban and Regional Planning, Abubakar Tafawa Balewa University, Bauchi, Nigeria; Tel: +2348023747882; E-mails: yybabanyara@gmail.com and yybabanyara@atbu.edu.ng

**Gabriella Marfe & Carla Di Stefano (Eds.)**

Healthcare waste means all the waste generated in medical facilities, laboratories, and centers of research associated with healthcare procedures. Additionally, it includes veterinary centers and waste generated during healthcare undertaken in the home (healing procedure) [5, 6]. Healthcare facilities mostly are situated in the urban centers as such; healthcare wastes that are not correctly managed can cause dangerous infection and pose a potential threat to the nearby environment, health workers (doctors, nurses, paramedical personnel) patients and the public [7]. It was established that between 75% to 90% of the waste generated by hospitals is comparable to domestic or municipal waste usually called "non- risks waste," general healthcare waste" or "non- hazardous waste", while "risks waste" or "hazardous waste" constitute the remaining 10% to 25% [8 - 10] when these wastes are mingled together they become hazardous to mankind, animals and the environment. Reports suggest that in developing countries 80% of healthcare waste is jumbled together with general or domestic waste [11]. Globally, it is estimated that per year about 7 to 10,000,000,000 tons of waste is generated, out of this only, 2,000,000,000 tons are general solid waste, of which medical waste contributes but a small fraction [11, 12]. These developing countries include Nigeria, Ghana, South Africa, India, and Pakistan to mention a few [10, 13]. Additionally, it has been documented that, globally, around 5.2,000,000 persons (including 4,000,000 children) die each year from diseases associated with the waste [2, 14]. Inadequate medical waste management causes land, water, and air pollution, gradual increase and augmentation of vectors such as worms, insects, and rodents and may cause the out break of diseases like cholera and malaria [15]. Furthermore, injuries from contaminated sharps, needles and syringes may lead to the transmission of hepatitis, and *acquired immunodeficiency syndrome* AIDS [15 - 17]. Medical waste workers should tackle the hazards associated with their practices, by segregating the healthcare waste at the source of generation [18, 19]. Poor attitudes and practices of medical waste management staff put them in danger. Healthcare waste workers in developing nations usually cultivated the poor attitude in performing their tasks right from inception and often becomes hard for them to adjust. For this reason, it is necessary to address these amidst medical waste workers (MWW) [18, 20]. Furthermore, medical waste (MW) is the second most dangerous waste after nuclear; this means trained healthcare personnel are needed to properly dispose of it. Therefore, a sound knowledge, attitude, and safe practice of healthcare personnel regarding the management of this waste are necessary [21, 22].

## PERSONNEL'S INVOLVED IN MEDICAL WASTE MANAGEMENT IN HEALTH CENTERS

According to the World Health Organization (WHO), medical waste workers most affected by medical waste include doctors, nurses, pharmacists, and other non-medical staff members. This is because they are routinely exposed to healthcare waste and risks from many fatal infections due to the indiscriminate management of waste. However, many of the affected healthcare workers are from third world countries where policies and systems to enforce management of health care waste are weak. It is approximated that more than five hundred healthcare workers lose their lives in sub-Saharan Africa yearly as a result of infection due to unsafe contact with medical waste. According to Nkonge Njagi A, *et al.*, [23] the personnel can be grouped into four cadres as follows:

a. **Medical practitioners:** this involves all categories of medical doctors, dentists, pharmacists, technologists, and clinical officers.
b. **Nurses:** this involves all categories of nurses in healthcare facilities.
c. **Laboratory technologists:** all cadres of technicians, technologists, and scientists are in this category.
d. **Hospital attendants:** this means all the attendants from the various wards and units including incinerator operators of the hospitals.

## IMPORTANCE OF KNOWLEDGE, ATTITUDE, AND PRACTICES IN THE MANAGEMENT OF HEALTHCARE WASTE

Inadequate knowledge, attitude, and practices of handling (managing) medical waste can lead to nonsocomial infection (Hospital-acquired infection). Definitions of these concepts as given by [23, 24] as follows:

a. **Knowledge:** clear awareness or explicit information regarding the hazards associated with medical waste must be made known to medical waste personnel and how to safely dispose of the waste is vital.
b. **Attitude:** this means peoples behavior, *e.g.* disseminating information regarding the positive attitude towards the environment and the protection of health is vital.
c. **Practice:** this concerns the behavior of healthcare personnel *e.g.* do they use protective equipment, are they following the rules and regulations of safe medical waste management.

## CLASSIFICATION OF HEALTHCARE WASTES

Medical waste is categorized into different types, namely: infectious waste, pathological waste, sharps, pharmaceutical waste, chemical waste, radioactive waste, cytotoxic agents and human or anatomical waste. Infectious waste contains pathogens (bacteria, viruses, parasites, or fungi) in sufficient concentration capable of transmitting a disease. Pathological waste consists of tissues, body organs or parts, and, body fluids. Sharps are objects that could cause cuts, or prick injuries, which may or may not be infected, such as, saws, needles, scalpels, broken glasses, *etc.* Pharmaceutical waste includes pharmaceutical products, drugs, and medicinal chemicals that are expired, spilled, or returned from patient wards, or are to be discarded because they are no longer needed. The chemical waste consists of solid, liquid, and gaseous chemicals to be discarded because they are no longer required. For instance from laboratories diagnostic or experimental work, or cleaning, housekeeping or disinfecting procedures. Chemical waste may be risks or non-risks wastes [25]. Radioactive waste is waste containing radioactive substances including solid, liquid and gaseous waste contaminated with a radionuclide [26]. Cytotoxic agents are substances that have a deleterious effect upon cells, used to cure cancer, for example, chemotherapy. Pressurized containers consisting of gas cylinders, cartridges, and aerosol cans that could burst if incinerated or accidentally punctured [26]. This can be presented diagrammatically below Fig. (**1**)

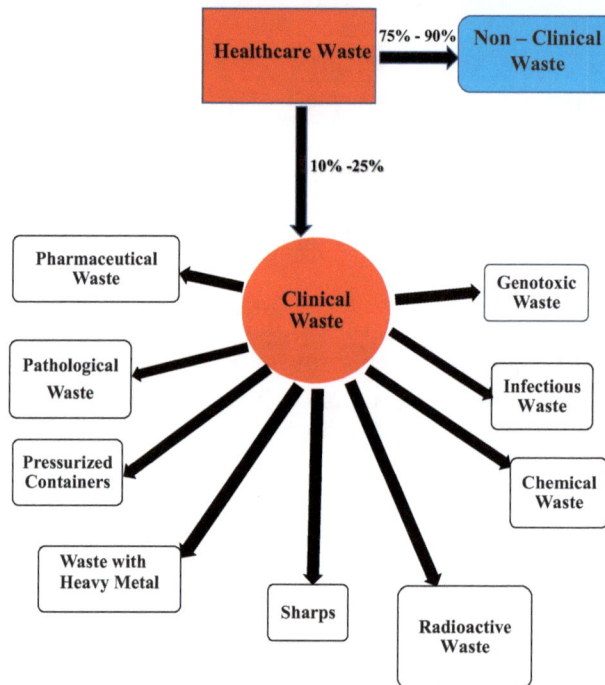

**Fig. (1).** Classification of Waste from Healthcare Facilities Adapted from [5, 17].

## Health and Risks Associated with Healthcare Waste Management

Most of the third world and underdeveloped cities in India, Nigeria, Ghana, and Pakistan, lack standard patterns or procedures for the safe management of medical waste. Medical wastes are disposed of indiscriminately, some were dumped in drainages and can be seen flowing smoothly *via* open drainage system up to the outskirts of towns and cities [27]. Below are some of the risks associated with poor medical waste management, *VIZ*:

a. Injuries from sharps to all categories of hospital staffs in and out of the surrounding health center, for instance, a person with sharp injuries from a contaminated needle is at risk of contracting hepatitis B (HBV) virus, hepatitis C (HCV), and Human immunodeficiency virus (HIV) which leads to acquired immune deficiency syndrome (AIDS). For instance, in 2010, injections with contaminated needles infected 33,800 people with HIV, 1.7 million people with (HBV), and 315,000 people with HCV [18].

b. Infection control that is poorly performed tends to super infects patients receiving treatment with (HIV, hepatitis B) especially those with poor immunity levels.

c. Healthcare waste personnel are exposed to the risks (hazards) of chemicals and drugs or pharmaceuticals. For instance, some chemicals and drugs (Vials, bottles) used in healthcare centers are hazardous. These substances are found in small quantities of medical waste. These may cause intoxication, injuries, and burns from the absorption of chemicals or pharmaceuticals *via*, inhalation, ingestion, mucous membrane and through the skin.

d. Persons not restrained by moral or ethical principles due to poor medical waste management could recycle disposables for repackaging and re-selling, this action is fatal.

e. Micro-organisms that are resistant or difficult to cure could develop.

f. The organic part of medical waste provides a breeding place for flies. The consequence is that flies can transmit diseases because they feed on filthy matter and food for humans.

g. Similarly, incineration which is the last stage of disposing of hazardous waste could infect the operators with the disease. According to the research done by Babanyara YY *et al.*, [17], a healthcare waste incinerator emits a large amount of toxic gases such as Dioxin and Furans and these are dangerous to health.

Due to the health risks associated with poor management of medical waste, the proper disposal and management of healthcare wastes are vital.

Examples of infections due to exposure to medical waste, with their medium of transmission, are shown in Table **1**.

**Table 1. Communicable diseases caused by exposure to medical wastes, causative virus, or bacterium and the transmission medium.**

| Type of infection | Causative virus or bacterium | Transmission medium |
|---|---|---|
| Gastroenteric infections | Enterobacteria, *e.g.Salmonella, Shigella* spp., *Vibrio cholerae, Clostridium difficile,* helminths | Faeces and/or vomit |
| Respiratory infections | *Mycobacterium tuberculosis,* measles virus, *Streptococcus pneumoniae,* severe acute respiratory syndrome (SARS) | Inhaled secretions, saliva |

*(Table 1) cont.....*

| Type of infection | Causative virus or bacterium | Transmission medium |
|---|---|---|
| Ocular infection | Herpesvirus | Eye secretions |
| Genital infections | *Neisseria gonorrhoeae*, herpesvirus | Genital secretions |
| Skin infections | *Streptococcus* spp. | Pus |
| Anthrax | *Bacillus anthracis* | Skin secretions |
| Meningitis | *Neisseria meningitidis* | Cerebrospinal fluid |
| Acquired immunodeficiency syndrome (AIDS) | Human immunodeficiency virus (HIV) | Blood, sexual secretions, body fluids |
| Haemorrhagic fevers | Junin, Lassa, Ebola and Marburg viruses | All bloody products and secretions |
| Septicaemia | *Staphylococcus* spp. | Blood |
| Bacteraemia | Coagulase-negative *Staphylococcus* spp. (including methicillian-resistant *S. aureus*), *Enterobacter*, *Enterococcus*, *Klebsiella* and *Streptococcus* spp. | Nasal secretion, skin contact |
| Candidaemia | *Candida albicans* | Blood |
| Viral hepatitis A | Hepatitis A virus | Faeces |
| Viral hepatitis B and C | Hepatitis B and C viruses | Blood and body fluids |
| Avian influenza | H5N1 virus | Blood, faeces |

## MICRO-ORGANISMS ASSOCIATED WITH HEALTHCARE WASTE

Different kinds of micro-organisms, *i.e.* viruses, protozoa, and bacteria that are capable of causing diseases have been examined by both cultivations and by (RT)-PCR assays. Many disease-causing bacteria, such as *Pseudomonas spp., Lactobacillus spp., Staphylococcus spp., Micrococcus spp., Kocuria spp., Brevibacillus spp., Microbacterium oxydans, and Propionibacterium acnes,* were recognized and documented from the various medical wastes. Also, disease-causing viruses such as noroviruses and hepatitis B virus have been identified in man's tissue wastes. Commonly detected bacterial and disease-causing micro-organisms such as *Pseudomonas spp., Corynebacterium diphtheriae, Escherichia coli, Staphylococcus spp.,* and respiratory synctial virus (RSV) have been documented to be part of the healthcare wastes. Healthcare waste must be controlled, monitored, and safely disposed of to prevent nosocomial infection due to the hazards of these wastes [28]. Medical waste leads to disagreement about its merits to human, animal and environmental health [28]. Manifestations of medically relevant bacteria in mounds of medical waste within a sanitary landfill and their antimicrobial susceptibility profile have also been investigated in the past by [28]. It also, documented that aliquots of leachate from medical waste in Brazil contained disease-causing bacteria *i.e.* strand of *Staphylococcus sp,* gram-negative rods of the *Enterobacteriaceae* family and non-fermenters. Bacterial resistance to all the antimicrobials tested was observed in all microbial groups, including resistance to more than one drug. This implies that bacteria in medical waste represent hazards to human and animal wellbeing. In the same vein, occurrences of multi-resistant strains prove the hypothesis that medical waste acts as a reservoir for resistance bacteria, with an environmental impact. The lack of regional laws about sorting, treatment and final disposal of waste may expose people to the hazards of contracting infectious diseases related to multi-resistant micro-organisms (See Table **2** below).

Table 2. Microbial diseases associated with Healthcare waste.

| Microbial Group | Type of Disease Caused |
|---|---|
| Bacteria | Tetanus, gas gangrene and other wound infection, anthrax, cholera, other diarrheal diseases, enteric fever, shigellosis, plague, *etc.* |
| Viral | Various Hepatitis, Poliomyelitis, HIV-infections, HBV, TB, STD rabies, *etc.* |
| Parasitic | *Amoebiasis, Giardiasis, Ascariasis, Ancylomastomiasis,Taeniasis, Echinococcosis, Malaria, Leishmaniasis, Filariasis,etc.* |
| Fungal infections | Various fungal infections like *Candidiasis, Cryptococcoses, Coccidiodomycosis,etc.* |

Adapted from [14, 17].

## HAZARDS OF IMPROPER DISPOSAL OF HEALTHCARE WASTE

Diseases outbreak as a result of infectious medical waste is inevitable unless the waste is properly treated in a manner that destroys the disease-causing organisms. Dangerous quantities of microscopic disease-causing organisms—viruses, bacteria, parasites or fungi—will be present in the waste if it is not handled properly. These micro-organisms can infect a person *via* gaps in the skin, mucous membranes, *via* inhalation, swallowing, or transmitted by vector organisms [29]. Human beings exposed to healthcare wastes are at risk greatly. For instance, medical management waste personnel, patients, visitors, scavengers, junkies, and those that are using're-used' contaminated syringes and needles. Additionally, Medical waste staff, waste pickers can become infected with HIV/AIDS and hepatitis B and C viruses *via* puncture of the skin or recycle of syringes/needles. These infections are dangerous [30]. Infectious stools or bodily fluids that are not treated before discharging into the river contaminates the water, this can lead to the outbreak of diseases. This practice is rampant in the third world.

For instance, no proper sterilization procedures are available in most cities in Africa, hence, the severity and size of cholera outbreaks in most parts of the African continent.

## CHEMICAL AND TOXIC THREATS

Chemical and pharmaceutical wastes in large volumes posed a threat to human health, animals, and the environment. These risks chemical wastes contain, toxic, corrosive, flammable, reactive, and/or explosive, and they can poison, burn or damage the skin and flesh of people who touch, inhale or are near to them. If burned, they may explode or produce toxic fumes. Some pharmaceuticals are toxic as well [30].

When chemical and pharmaceutical waste is disposed of in unlined landfills, especially unlined pits, these wastes may contaminate ground and surface water—particularly when large quantities are disposed of. This can threaten people who use the water for drinking, bathing and cooking, and damaging plants and animals in the local ecosystem. Burning or incinerating healthcare waste, while often a better option than disposal in an unlined pit, may create additional problems. Burning or incineration of healthcare waste may produce toxic air pollutants such as Nitrogen Oxides (NOx), particulates, dioxins, and heavy metals and distribute them over a wide area. Dioxins and heavy metals are of particular concern [31]. Dioxins believed to be potent cancer-causing agents, do not biodegrade and accumulate in progressively higher concentrations as they move up the food chain [32].

Heavy metals such as mercury and cadmium are toxic and/or cause birth defects in small quantities and can also concentrate on the food chain. Disposable pressurized containers pose another hazard for incineration, as they can explode if burned.

The disposal of large quantities of hazardous chemicals and pharmaceuticals is a serious problem. In most of Africa, no methods are available to small-scale facilities that are safe and affordable [31].

### Types of Chemicals Common to Waste

*Mercury*

Mercury is considered a heavy metal occurring naturally. It is a silvery-white liquid that readily vaporizes at room temperature and pressure and may stay in the atmosphere for up to a year. Mercury is transported by wind currents when released to the environment and ultimately accumulating in marine and lake as sediments. In the marine environments, inorganic mercury compound can be transformed by bacteria into organic form methyl mercury which accumulates in fish tissue which when ingested, affects human health.

It is important to note that mercury is highly toxic in elemental form or as methyl mercury. It may be harmful if absorbed through the skin and fatal if inhaled. 80% of the inhaled mercury vapor is absorbed into the blood *via* the lungs. This harmed nervous, digestive, respiratory, immune systems and kidneys, as well as the lungs. Health issues from mercury exposure can be tremors, impaired vision and hearing, paralysis, insomnia, emotional instability, developmental deficits during fetal development, and attention deficit and developmental delays during childhood. Mercury may have no threshold below which some adverse effects occur as recent studies show [24]. Several medical devices use Mercury, especially fever thermometers and blood-pressure monitoring equipment. This will be a hazard in terms of breakage and disposal. Batteries are a well-known source of mercury in health care waste, especially the small button batteries. Manufacturers from America and Europe are removing mercury from their products, but it is still present in those produced elsewhere [33]. Recently, health-care institutions have adopted a policy of gradual replacement with mercury-free alternatives.

Health-care institutions contribute up to 5% of the mercury release to water bodies through untreated wastewater. It was estimated by Environment Canada that one-third of mercury load in sewerage systems comes from dental practices.

Medical waste incineration is considered as one of the main sources of mercury released into the environment from health-care institutions. According to the United States, Environmental Protection Agency medical incinerators may have historically contributed up to 10% of mercury air releases. Also, mercury contained in dental amalgam, laboratory and other medical devices contributes more than 50% of total mercury emissions in the United Kingdom [24, 34].

*Silver*

In recent times, there is a decrease in the use of mercury in health care. But another toxic heavy metal, silver, is being used as a bactericide and in nanotechnology. When silver is used in large doses, it can turn a person's skin to permanently grey. Also, bacterial development of resistance to the metal and subsequent development of resistance to antibiotics are the increasing concern with both regulators and others about the potential effects of silver [35, 36].

## *Disinfectants*

Chlorine and quaternary ammonium disinfectants are used in large quantities in our health-care institutions, but are corrosive. These reactive chemicals may form highly toxic secondary compounds. When chlorine is used in a place without ventilation, it generates chlorine gas as a by-product of its reaction with organic compounds [5, 24]. Therefore, it important to use good working practices to avoid creating conditions where the concentrations in the air exceed safety limits.

## *Pesticides*

This type of chemical found in waste when outdated, stored in leaking drums or torn bags, can directly or indirectly affect human health when in contact with them. Ground waters are contaminated with pesticides when it seeps into the ground during heavy rains. This poisons humans through direct contact with a pesticide formulation, inhalation of vapors, drinking or eating contaminated water or food. Spontaneous combustion of improperly stored and contamination as a result of inadequate disposals, such as open burning or indiscriminate burying may be hazardous [29].

## GENOTOXIC WASTE HAZARDS

Genotoxic wastes are substances from certain cytotoxic drugs or vomit, urine, and feces from patients treated with cytotoxic drugs, chemicals, and radioactive materials. Special care in handling the waste is essential because of the severity of the hazards for health-care workers responsible for handling or disposal. It is imperative to note that genotoxic waste is governed by a combination of the substance which is toxic itself and depends on the extent and duration of exposure. During the preparation of, or treatment with, particular drugs or chemicals genotoxic substances may be exposed which threatens health. The waste can go into a human through inhalation of dust or aerosols, absorption through the skin, ingestion of food accidentally contaminated with cytotoxic drugs, and ingestion as a result of bad practice, such as mouth pipetting, or from waste items. It may also occur through contact with body fluids and secretions of patients undergoing chemotherapy.

The cytotoxicity of many antineoplastic drugs is cell-cycle specific, targeted on specific intracellular processes such as DNA synthesis and mitosis. Others such as alkylating agents are not cell-cycle specific but are cytotoxic at any point in the cell cycle. Studies conducted experimentally by [29, 32] have shown that many antineoplastic drugs are carcinogenic and mutagenic which has secondary neoplasia, that is, occurring after original cancer has been eradicated, is known to be associated with chemotherapy.

Direct contact with skin or eyes with many cytotoxic drugs makes extreme irritations that have harmful local effects. Dizziness, nausea, headache or dermatitis may also be caused by cytotoxic drugs. Discharge of genotoxic waste into the environment could have disastrous ecological consequences.

## RADIOACTIVE WASTE AND ITS HAZARDOUS NATURE

The type and extent of exposure are the determinants of the nature of illness caused by radioactive waste, which can range from headache, dizziness, and vomiting to more serious problems. Radioactive waste in a sufficiently high radiation dose may affect genetic material. Highly active radioactive sources, such as those used in diagnostic instruments (*e.g.* gallium sealed sources) may cause severe injuries, including tissue destruction which may necessitate the amputation of the body.

Contamination of external surfaces of containers or improper mode or duration of waste storage is the hazard of low-activity radioactive waste. Medical health workers, and waste handling and cleaning personnel exposed to radioactivity are mostly at very high risk [29, 37].

## ANTIBIOTIC RESISTANCE THAT IS WIDELY SPREAD IN THE ENVIRONMENT DUE TO INDISCRIMINATE MEDICAL WASTE DISPOSAL

When microorganisms resist an antimicrobial medicine to which it was previously sensitive, such microorganisms possess Antimicrobial resistance (AMR). Resistant organisms include bacteria, viruses, and some parasites which can withstand attacks by antimicrobial medicines, such as antibiotics, antiviral, and anti-malaria so that treatments become ineffective and infections persist which may spread to others. The consequences of the use and the misuse of antimicrobial medicines give rise to AMR which develops when a microorganism acquires a resistance gene [37]. A microbial pathogen that resists at least three groups of antibiotics is described as a multidrug resistance pathogen. Medical liquid wastes are the reservoirs of harmful infectious agents such as the pathogens and multiple drug-resistant microorganisms [38]. Infectious hazards of medical waste include the spread of infectious diseases and microbial resistance from healthcare institutions into the environment thereby posing risks of getting infections and antibiotic resistance in the communities [38]. Water system and the stream are contaminated when a hospital discharges medical liquid waste into the sewage system without proper treatment, such concentrated forms of infectious agents and antibiotic resistant-microbes are shed into communities resulting in water borne diseases such as cholera, typhoid fever, dysentery, and gastroenteritis. Wastewater, surface water, groundwater, sediments, soils, and other environmental compartments have been found to contain antibiotics, disinfectants, and bacteria-resistant to them [39]. Studies have discovered traced level concentrations of antibiotics in wastewater treatment plant effluents and surface waters [40]. When microorganisms exposed for long term to low concentrations of antibiotics in wastewater and surface water, there is the potential of the development of antibiotic resistance in these organisms [41].

An increase in antimicrobial resistance is of great concern globally. In the United Kingdom, a report stated that the resistance to antibiotics and other anti-infective agents that constitute a major threat to public health ought to be recognized. Resistant bacteria and antibiotics, when disposed into the environment, can disturb the established well balanced and important system [42]. The input of resistant bacteria into the environment seems to be a source of great concern to waste managers.

Also, the development of antibiotic-resistance in bacteria and their disposal in the environment is of serious public health concern because an individual patient can develop an antibiotic-resistant infection by contacting a resistant organism and spread in the communities.

Hospitals and other public health institutions must safeguard the health of the community. But the waste produced by them if disposed of improperly can pose an even greater threat than the original diseases themselves due to the presence of concentrated forms of risks including pathogenic and antibiotic-resistant microorganisms [38]. It was reported in Nepal that several thousand die due to infectious diseases and several more, losing the quality of lives [38]. And untreated hospital liquid waste discharge into surface water directly or indirectly must have been adding more health problems to our society. It is our common observation that the majority of healthcare institutions do not practice safe healthcare liquid waste management.

## HEALTHCARE WASTE MANAGEMENT PRACTICE (S) IN DEVELOPING COUNTRIES CASE STUDIES

Several studies regarding healthcare waste suggest that poor management of healthcare waste is harmful to the health of the patients, health workers, and the environment. Waste produced during medical care activities in Kenya led to health complications such as respiratory, reproductive and the nervous system of the persons exposed to it [43].

A research conducted in the Mazandaran province in Iran reported that lack of knowledge and a low understanding of universal precautions on healthcare workers was the contributing factor to poor medical waste management [44]. In the same vein, a study carried out in India documented that lack of knowledge contributed to poor waste management practices and it was suggested that there was a need to develop waste management protocol for health and to develop waste management audits after realizing that waste management was generally poor [45]. Furthermore, a survey with regards to the attitudes and knowledge of pharmacists regarding medical waste during an educational campaign for proper medication and pharmaceutical waste disposal found out that, a brief educational intervention was effective in changing the attitude and practices of pharmacists. Pharmacists that received education on proper disposal improved their attitudes than those who did not receive any education at all [46].

According to the research conducted by Rhadar R., [47] in India, biomedical waste management amongst healthcare workers in tertiary care rural hospitals supported the idea that adequate knowledge was the first step towards developing favorable attitudes to practice waste management effectively. Health workers with adequate knowledge were better in terms of practice than those with less knowledge.

In a study conducted at Northwest Cameroon by Mochungong PI, [48], it was posited that the health hazards associated with medical waste and unsafe disposal practices have brought concerns regarding its impact on the community's health. The study further suggested that inadequate or lack of knowledge of safe medical waste disposal practices contributed to poor disposal practices. According to the study conducted by Mochungong PI, [48], 47.5% of healthcare wastes workers (HCWs) were not aware of any government document internal or external on safe disposal of medical waste [48].

Furthermore, a survey carried out in Tanzania between 2003 and 2005 with regards to healthcare waste management systems reported that the level of knowledge on medical waste management amongst healthcare workers was low and training on waste management was needed. The survey documented that the rapid growth of populations and poor medical waste management system lead to an increase in the use of disposables materials and thus, the main factor leading to the accumulation of a large amount of healthcare waste in the medical facilities. Additionally, most of the health centers in Tanzania had low incineration capacity with few of them constructed with bricks [49]. Furthermore, garbage pit meant for the burial of garbage was converted for the disposal of the used sharps and syringes and these pits were shallow scavengers could easily retrieve them for reuse. The study recommended that health education was important to improve medical waste practices [49].

A research done in the urban area of Karachi Pakistan regarding healthcare waste personnel reported that general waste and infectious waste were mingled together and disposed of in landfills; liquid waste was also not treated before discharging into the environment, therefore, the study concluded that knowledge, attitude, and practices of healthcare workers was extremely poor and proper facilities for the management of hospital waste was almost non-existent in Karachi [50]. Another study found that the nurses that were impacted knowledge about safe medical waste management, had satisfactory practices about clinical waste management but their level of understanding and practices varies [51].

In Ghana, a study done by Squire NJ [52], documented that most of the medical laboratories in Accra Ghana were not aware of the rules and regulations about sorting or segregation laws. Only a few practiced separations of their waste before disposal and liquid waste was discharged into the environment without treatment which in turn could pollute the surface water and the underground water systems. There is absence of bio-hazardous signs on the vehicles for the transportation of healthcare waste and the drivers had no training.

In Algarve hospitals (in Portugal, a developed country) based on the research done by Fereira V, [53] they reported that there was a lack of training in hospitals due to lack of resources, but nurses showed higher knowledge on medical waste management than doctors and the medical staff showed a low perception of risk associated with infectious waste. Needle stick injuries according to the study were attributed to poor segregation and containerization of waste materials. Other authors in their study [54] attributed poor waste management to medical professionals towards lack of commitment and accountability in ward management because nurses had a better understanding and more responsible for implementation. In India, due to insufficient knowledge of the correct method of medical waste disposal in Kanchipuram town India, there was a need to develop programs that not only give knowledge to doctors but also to motivate them to actively practice proper biomedical waste management [55]. However, the survey conducted in the northern part of Jordan showed healthcare waste management personnel in Irbid city performed poor practices when it comes to handling, storage, and disposal of waste generated in comparison to developed countries [56]. Knowledge and training on medical waste management were described as inadequate in Bangladeshi where only 23 Bangladeshi waste workers have received the most basic information from non-governmental organizations than their employers.

The challenge in good medical waste management was also found in a Palestinian hospital in the West Bank where there was inadequate segregation between the risks medical waste and non-risk waste; this is attributed to lack of standard rules guiding medical waste management and thus, resulted in dumping hospital waste with the municipal waste *i.e.* (Co-mingling). The study further reported that medical waste management staffs were exposed to health hazards due to the absence of protective equipment and formal training of managing healthcare waste [57].

A study conducted regarding medical waste disposal at Yenagoa, South Nigeria in 2011 revealed that standard practice is not adhered to in the disposal of medical waste. About 2,000,000 kg of medical waste is generated yearly with infectious wastes and sharp items constituting nearly 19%. These wastes are disposed of in public waste bins along streets and often dumped in open and unlined landfills [58].

Another study that was conducted in Gondor Town, Northwest of Ethiopia, on factors associated with the risk perception of medical waste personnel towards medical care waste management reported that healthcare waste in Ethiopia is treated equally like any other municipal waste, no health care facilities had healthcare waste management guidelines only small quantities of medical waste personnel had medical waste management guidelines. The study recommended that in-service training on medical waste must be taken to improve the perception of medical waste management personnel and hospitals must have waste management facilities and guidelines for effective medical waste management [59].

Several studies in Brazil with regards to medical waste management indicated that health and environmental risk associated with medical waste can be reduced by setting a standard procedure and involving waste management personnel [60]. The study conducted by Squire NJ [52], in India showed that knowledge was a detrimental tool to good practices. When compared to knowledge with attitude and practice the personnel with higher education such as the consultants, residents, and scientists had very good knowledge but a low percentage of this category of people has the same kind of attitude and practice.

The management of medical waste in several hospitals in the third world is ineffective as reported by Hossain MS, *et al.*, [61]. Due to lack of safety equipment, absence of training of the waste management staff and these exposed the health care waste personnel to health risks; pollution of the environment is inevitable as a result of the risks associated with biomedical waste. The study suggested that the main reasons for the mismanagement of medical solid waste were due to lack of rules and regulations, lack of finance, lack of technical staff to handle the management of the medical waste [61].

An audit on waste management at Gauteng province South Africa conducted by Fischer D [62], suggested that excessive and incorrect manual handling of medical waste, unsafe utilization of equipment and the excessive emission of pollutants from healthcare risk waste treatment plants were identified to cause problems in the medical waste management.

Additionally, research done in Abuja Nigeria with regards to the medical waste disposal technique used in the management of solid waste, reported that health workers, environment, and patients are at risk of being infected from sharps and pollution with nonsocomial diseases due to improper disposal of the healthcare waste. The results showed that the average waste per bed was 2,78kg of healthcare waste and 26, 5% of the waste was hazardous. The study concluded that waste management personnel do not have formal training in healthcare waste management techniques and hospital administrators don't impose sound management of medical waste. The study also recommended that waste generators must be educated to manage their waste so that their patients and environment are protected. The study argued that a hospital provides medical services with no emphasis on medical waste as a by-product [63].

Based on the study done by Phenxay S, *et al.*, [64] in two selected hospitals in the Laos Democratic Republic healthcare waste management is poor; sorting of the waste is absent or highly improper segregation of healthcare waste is done at both the primary and secondary hospitals. Additionally, a large volume of healthcare waste was generated from the inpatient unit of the primary health care facility and thus, urgent attention and enlightenment are needed. In their study [65] reported that audit of waste disposal practice of ten dental clinics was improperly done due to a lack of regulations or policy to guide dental waste management. Furthermore, the improper practice of medical waste was evident on the study conducted by Bansal M [66], amongst dental practitioners in Tricity (Mohali, Panchkula, and Chandira) where 14% of the personnel doesn't know the types of waste generated at the dental clinics and 32% of the dentists dispose of dental waste inappropriately. Finally [67], suggested that medical waste personnel should be responsible for the proper management of medical waste and intensive knowledge in the management of clinical waste is vital and in turn, will lead to proper practice.

## RESULTS

The findings from the reviewed articles revealed the following: lack of safety equipment and immunization in most of the health centers, inadequate or lack of awareness about the health risks associated with medical waste, lack of knowledge/training with regards to safe medical waste management, absence of waste management and disposal systems, lack of funds and technical staff as a result of the low priority given to medical waste management. Most developing nations don't have standards, laws, and regulations for safe medical waste management and where the laws exists the countries do not enforce them. An important aspect of the management is ascribing responsibility for the handling and disposal of medical waste to the producer. Polluter pays principle (PPP) States that the responsibility lays with the waste producer, usually the healthcare provider, or the hospital involved in medical care activities. For safe and sustainable management of medical waste, financial analyses must include all the costs of disposal.

## RECOMMENDATIONS/CONCLUSION

Sound medical waste management relies on these distinct groups:

a. Creating a complete and reliable system, assigning duties, resource allocation, handling, and disposal. This is a long-term process, sustained by gradual improvements;
b. Disseminating of information about the hazards associated with medical waste and the merits of following a standard procedure for safe and sound practices;
c. Choosing a safe and environmentally-friendly management option, to protect people from risks when collecting, handling, storing, transporting, treating or disposing of healthcare waste;
d. Strong political will by the Governments is necessary to achieve long-term improvement, although immediate action can be taken locally;

In conclusion, if the above recommendations would be taking into considerations, it will go a long way in ameliorating the problems with regards to poor management of healthcare waste in most health centers of the third world.

## CONSENT FOR PUBLICATION

Not applicable.

## CONFLICT OF INTEREST

The authors confirm that the contents of this chapter have no conflict of interest.

## ACKNOWLEDGEMENTS

Declare none.

# REFERENCES

[1]     Mathur V, Dwivedi S, Hassan MA, Mishra R. Practices about biomedical waste management among healthcare personnel: A cross-sectional study. Ind J Commun Med 2009; 36(2): 143-5.
[PMID: 21976801]

[2]     Musa M. Knowledge, Attitude, and Practices of Healthcare Workers on Medical Waste Management in Jama'are General Hospital Bauchi, Nigeria" Unpublished BTech Thesis Abubakar Tafawa Balewa University, Bauchi, Nigeria 2019.

[3]     Tabrizi JS, Saadati M, Heydari M, Rezapour R, Zamanpour R. Medical waste management improvement in community health centers: an interventional study in Iran. Primary Health Care Res Devel 2018; 1-6.
[http://dx.doi.org/10.1017/S1463423618000622]

[4]     Chowdhary A. Study of knowledge, behaviour and practice of biomedical waste among health personnel. Int J Commun Med Public Health 2018; 5(8): 3330-4.

[5]     Babanyara YY. An assessment of medical waste management practice(s) of Ahmadu bello university teaching hospital (ABUTH) Zaria Nigeria" Unpublished MTech Thesis Abubakar Tafawa Balewa University, Bauchi, Nigeria 2012.

[6]     Teshiwal D, Hassen F, Kasaw A, Aster T. Assessment of knowledge, attitude, and practice about biomedical waste management and associated factors among the healthcare professionals at debre markos town healthcare facilities, northwest ethiopia. Hindawi J Environ Public Health 2018; 2018: 10.
[http://dx.doi.org/http://doi.org/10.1155/2018/7672981] [PMID: 7672981]

[7]     Safe management of wastes from health-care activities.In: Bulletin of the World Health Organization. 2014.

[8]     Chartier Y, Emmanuel J, Pieper U, *et al.* 2014. http://www.searo.who.int/srilanka/documents/safe_management_of_wastes_ from_healthcare_activities.pdf?ua=1

[9]     Babanyara YY, Gana BA, Garba T, Batari MA. Environmental and Health risks Associated with Dental Waste Management: A Review 2015.www.iiste.org

[10]    Khan BA, Cheng L, Khan AA, Ahmed H. Healthcare waste management in Asian developing countries: A mini review. Waste Manag Res 2019; 37(9): 863-75.journals.sagepub.com/home/wmr
[PMID: 31266407]

[11]    Mugabi B, Hattingh S. Assessing knowledge, attitudes, and practices of healthcare workers regarding medical waste management at a tertiary hospital in botswana: a cross-sectional quantitative study. Niger J Clin Pract 2018.www.njcponline.com

[12]    Singh VP, Biswas G, Sharma JJ. Biomedical waste management – An emerging concern in Indian hospitals. Ind J Forensic Med Toxicol 2007; 1: 39-44.

[13]    Kumar R, Somrongthong R, Ahmed J. Effect of medical waste management trainings on behavior change among doctors *versus* nurses and paramedical staff in Pakistan. J Ayub Med Coll Abbottabad 2016; 28(3): 493-6.
[PMID: 28712220]

[14]    Akter N. Medical Waste Management: A Review. Thailand: Asian Institute of Technology, School of Environment, Resources and Development 2000.

[15]    Shehu RY. Knowledge, Attitude, and Practices of Healthcare Workers on Bio-Medical Waste Management in New Specialist Hospital Bauchi, Nigeria" Unpublished BTech Thesis Abubakar Tafawa Balewa University, Bauchi, Nigeria 2018.

[16]    Mathur V, Dwivedi S, Hassan M, Misra R. Knowledge, attitude, and practices about biomedical waste management among healthcare personnel: a cross-sectional studyt. Indian J Community Med 2011; 36(2): 143-5.

[PMID: 21976801]

[17]     Babanyara YY, Ibrahim DB, Garba T, Bogoro AG, Abubakar MY. Poor Medical Waste Management (MWM) practices and its risks to human health and the environment: a literature review. J World Acad Sci Eng Technol 2013; 7(11): 801-8.

[18]     Ahmed AH, Terry T, Mentore V. Healthcare waste management: A case study from sudan. J of Environts 2018.www.mdpi.com/journal/environments

[19]     McVeigh P. Nursing and environmental ethics, medical waste reduction, reuse and recycling. Today's OR Nurse 1993; 15(1): 13-8.
         [PMID: 8456452]

[20]     Pinto VN, Joshi SM, Velankar DH, Mankar MJ, Bakshi H, Nalgundwar A. comparative study of knowledge and attitudes regarding biomedical waste (BMW) management with a preliminary intervention in an academic hospital. Int J Med Public Health 2014; 4(1): 91-5.

[21]     Madhukumar S, Ramesh G. Study about awareness and practice about healthcare waste management among staff in a medical college hospital, Bangalore. Iran J Basic Med Sci 2012; 3(1): 7-11.

[22]     Dzekashu LG, Akoacher JF. Mbacham, WF Medical waste management and disposal practices of health facilities in kumbo east and kumbo west health districts. Int J Med Sci 2017; 9(1): 1-11.
         [PMID: 28138303]

[23]     Nkonge Njagi A, Mayabi Oloo A, Kithinji J, Magambo Kithinji J. Knowledge, attitude, and practice of healthcare waste management and associated health risk in the two teachings and referral hospital in Kenya. J Commun Health 2012; 37: 1172-1177.WHO.

[24]     Mercury in health care policy paper. Geneva, World Health Organization 2005. http://www.who.int/water_sanitation_health/medicalwaste/mercury polpap230506.pdf

[25]     Malebatja SM. Knowledge and Practices of Health Care Workers on Medical Waste disposal at George Masebe Hospital, Waterberg District, Limpopo Province, South Africa" Unpublished, MSC Thesis University of Limpopo, South Africa 2016.

[26]     Chartier Y, Emmanuel J, Pieper U, Pruss A, Rushbrook P, Stringer R. Safe management of wastes from healthcare activities. 2nd ed., Switzerland: World Health Organization (WHO) Press 2014.

[27]     Pandey S, Dwivedi AK. Nosocomial infections through hospital waste. Int J Waste Resource 2016; 6: 200.
         [http://dx.doi.org/10.4172/2252-5211.1000200]

[28]     Nascimento TC, Januzzi WdeA, Leonel M, Silva VL, Diniz CG. [Occurrence of clinically relevant bacteria in health service waste in a Brazilian sanitary landfill and antimicrobial susceptibility profile]. Rev Soc Bras Med Trop 2009; 42(4): 415-9.
         [PMID: 19802478]

[29]     1992.http://whqlibdoc.who.int/hq/1994/ WHO_PEP_RUD_94.1.pdf

[30]     Johannessen LM, Dijkman M, Bartone C, Hanrahan D, Boyer G, Chandra C. Health care waste management guidance note, Health Nutrition and Population discussion paper. Washington, DC: The International Bank for Reconstruction and Development, The World Bank 2000.

[31]     Prüss A, Townend WK. Teacher's Guide- Management of wastes from health-care activities. Geneva: World Health Organization 1998. http://www.who.int/ environmental_information/ Information_resources/worddocs/HCteachguid/health_care_wastes_teacher.htm

[32]     WHO. Healthcare wastes 1999. http://www.who.int/water_sanitation_ health/Environmental _sanit/MHCWHanbook.htm

[33]     Directive 2006/66/EC of the European Parliament and of the Council of 6 September 2006 on batteries and accumulators and waste batteries and accumulators and repealing Directive 91/157/EEC. Off J Eur Union L 2006; 266: 1-13.

[34]   Risher JF. Elemental mercury and inorganic mercury compounds: human health effects 2003.http://www.who.int/ipcs/publications/cicad/en/cicad50.pdf

[35]   Chopra I. The increasing use of silver-based products as antimicrobial agents: a useful development or a cause for concern? J Antimicrob Chemother 2007; 59(4): 587-90.
[PMID: 17307768]

[36]   Senjen R, Illuminato I. Nano and biocidal silver 2009.http://www.foe. org/sites/default/files/Nano-silverReport_US.pdf

[37]   2012.Antimicrobial resistance http://www.who.int/mediacentre/ factsheets/fs194/en/

[38]   Sharma DR, Pradhan B, Mishra SK. Multiple drug resistance in bacterial isolates from liquid wastes generated in central hospitals of Nepal. Kathmandu Univ Med J (KUMJ) 2010; 8(29): 40-4.
[PMID: 21209506]

[39]   Kümmerer K. Resistance in the environment. J Antimicrob Chemother 2004; 54(2): 311-20.
[PMID: 15215223]

[40]   Kolpin DW, Furlong ET, Meyer MT, *et al.* Pharmaceuticals, hormones, and other organic wastewater contaminants in U.S. streams, 1999-2000: a national reconnaissance. Environ Sci Technol 2002; 36(6): 1202-11.
[PMID: 11944670]

[41]   Smith KE, Besser JM, Hedberg CW, *et al.* Quinolone-resistant Campylobacter jejuni infections in Minnesota, 1992-1998. N Engl J Med 1999; 340(20): 1525-32.
[PMID: 10332013]

[42]   Hiraishi A. Respiratory quinone profiles as tools for identifying different bacterial populations in activated sludge. J Gen Appl Microbiol 1998; 34: 39-56.

[43]   Kaseva ME, Mato RRAM. Critical review of industrial and medical waste practices in Dar es salaam city. Resour Conserv Recycling 1999; 25: 271-87.
[http://dx.doi.org/10.1016/S0921-3449(98)00068-8]

[44]   Khalilian A, Mahmood F, Motamed M. Knowledge, attitude and practices of health care workers and medical students towards universal precautions in hospitals in Mazandaran province. Mediterranean Health J 2006; 12(5): 653-61.

[45]   Mostafa GM, Shazly MM, Sherief WI. Development of a waste management protocol based on assessment of knowledge and practice of healthcare personnel in surgical departments. Waste Manag 2009; 29(1): 430-9.
[PMID: 18316184]

[46]   Jarvis CI, Seed SM, Silva M, Sullivan KM. Educational campaign for proper medication disposal. J Am Pharm Assoc (2003) 2009; 49(1): 65-8.
[PMID: 19196599]

[47]   Rhadar R. Assessment of the existing knowledge, attitude and practices regarding biomedical waste management among health care workers in tertiary care rural hospital. J Health Sci Res 2012; 7: 2249-9571.

[48]   Mochungong PI. The plight of clinical waste pickers: evidence from the Northwest region of Cameroon. J Occup Health 2010; 52(2): 142-5.
[PMID: 20110620]

[49]   Manyele SV, Anicetus H. Management of medical waste in Tanzanian hospitals. Tanzan Health Res Bull 2006; 8(3): 177-82.
[PMID: 18254511]

[50]   Sultana H, Salahudin A. Waste disposal of government health care facilities in urban area of Karachi: KAP survey 2007. Pak J Med Res 2007; 46(1): 2.

[51]   Yadavannavar M, Berad AS, Jagirdar P. Biomedical waste management: A study of knowledge, attitude, and practices in a tertiary health care institution in bijapur. Indian J Community Med 2010; 35(1): 170-1.
[PMID: 20606945]

[52]   Squire NJ. Biomedical pollutants and urban waste management in the Accra metropolitan Area, Ghana: a framework for urban management of the environment (FUME) Unpublished PhD Thesis, University of Waterloo, Ontario, Canada 2012.

[53]   Fereira V, Teixera MR. Assessing the medical waste management practices and associated risk perceptions in Algarve hospitals, Portugal. Waste Manag 2012; 30: 2657-63.

[54]   Saini S, Nagarajan SS, Sharma RK. Knowledge, attitude and practices of biomedical waste management amongst staff of a tertiary level hospital in India. J Acad Hosp Adm 2005; 17.

[55]   Selvaraj K, Sivaprakasam P, Sudhir BT, *et al.* Knowledge and practice of biomedical waste management among medical practitioners of Kanchipuram Town. Int J Curr Microbiol Appl Sci 2013; 2(10): 262-7.

[56]   Bdour A, Altrobshesh N, Hadadin N, Shereif AL. Assesment of medical waste management practice: A case study of the northern part of Jordan. Waste Manag 2007; 27(6): 75-746.

[57]   Al-Khatib IA, Khatib RA. [Assessment of medical waste management in a Palestinian hospital]. East Mediterr Health J 2006; 12(3-4): 359-71.
[PMID: 17037705]

[58]   Chima GN, Ezekwe IC. An assessment of medical waste management in health institutions in Yenagoa, South-South, Nigeria 2011 World Review of Science, Technology and Sustainable Development 2011; 224-33.

[59]   Muluken A, Haimanot G, Messafint M. Health care waste management practices among health care workers in health care facilities of Gondor town, North West Ethiopia. Health Sci J 2012; 7(3): 2012.

[60]   Da Silva CE, Hoppe AE, Ravanello MM, Mello N. Medical wastes management in the south of Brazil. Waste Manag 2005; 25(6): 600-5.
[PMID: 15993344]

[61]   Hossain MS, Santhanam A, Nik Norulaini NA, Omar AK. Clinical solid waste management practices and its impact on human health and environment--A review. Waste Manag 2011; 31(4): 754-66.
[PMID: 21186116]

[62]   Fischer D. Health Care Waste Management in Gauteng: Audit report.

[63]   Bassey BE, Benka-Coker MO, Aluyi HS. Characterization and management of solid medical wastes in the Federal Capital Territory, Abuja Nigeria. Afr Health Sci 2006; 6(1): 58-63.https://www.ncbi.nlm.nih.gov/pmc/articles
[PMID: 16615831]

[64]   Phenxay S, Miyoshi M, Salasaka K, Kurowa C, Kurma J. Health care waste management in Lao PDR: A case study. J Waste Manag Res 2005; 23: 571-81.

[65]   Glenda M, Stankiewitz N. Audit of waste collected over one week from ten dental practioners: A pilot study. Wiley online. Library (Lond) 2008;
[http://dx.doi.org/10.1111/j.1834-7819.1997.tb00106.x]

[66]   Bansal M, Vashisth S, Gupta N. Knowledge, awareness and practices of dental care waste management among private dental practitioners in Tricity (Chandigarh, Panchkula and Mohali). J Int Soc Prev Community Dent 2013; 3(2): 72-6.
[http://dx.doi.org/10.4103/2231-0762.122436] [PMID: 24778983]

[67]   Chaerul M, Tanaka M, Shekdar AV. A system dynamics approach for hospital waste management. Waste Manag 2008; 28(2): 442-9.
[PMID: 17368013]

**CHAPTER 3**

# Blood Exposure Accidents: Knowledge and Evaluation of Health Professional in the Emergency Pavilion of the Hospital of Batna City

**Sefouhi Linda*, BenBouza Amina** and **Houfani Roufaida**

*Natural Risks and Territory Planning Laboratory (LRNAT), Institute of Industrial Hygiene and Safety, Batna 2 University, Batna, Algeria*

**Abstract:** Blood Exposure Accidents (BEA) are a real risk to personnel health, particularly in developing countries. The objective of this study is to assess the knowledge and attitudes of health personnel on BEA, know the existing safety equipment and its application by the staff, and propose improvements to have an effective protection prevention system. We have, therefore, conducted a survey during the month of May 2019 in order to assess the current situation. A previously developed questionnaire was provided to health personnel who had direct daily contact with patients. The results showed a good knowledge of the risk and indicate that accidents involving exposure to blood are frequent and can result in serious consequences, including infection with hepatitis B and C viruses, as well as human immunodeficiency virus (HIV), they may be due to neglect of preventive measures, lack of contact with occupational medicine, work overload…*etc.* Blood Exposure Accidents are still a concern in our community. There is a high rate of needle stick injuries in daily hospital practice. Then information tools are needed both training for staff with regard to BEA, can reduce the severity of exposure.

**Keywords:** Accidents of blood exposure, Health personnel, Infection, Knowledge, Practice, Protection health, Risk.

## INTRODUCTION

Health establishment, like any organization, has a heavy responsibility towards the personnel it employs who are likely to be affected by an occupational accident or disease related to working conditions (presence of pathogens, use of sensitive technologies or devices, *etc.*).The risk of contracting the work-related illness is therefore high due to the use of toxic substances, exposure to infectious diseases or accidents of blood exposure.

---

* **Corresponding author SEFOUHI Linda:** Natural Risks and Territory Planning Laboratory (LRNAT), Institute of Industrial Hygiene and Safety, Batna 2 University, Batna, Algeria; Emails: lsefouhi@yahoo.fr & l.seffouhi@univ-batna2.dz

**Gabriella Marfe & Carla Di Stefano (Eds.)**

Blood Exposure Accidents (BEA) involve a risk of transmission of infectious agents and concern all germs carried by the blood. Many bacterial, viral, and parasitic infections can be transmitted to the personnel. Healthcare workers (HCWs) are exposed to blood and body fluids due to occupational accidents, which can result from percutaneous injury (needle stick or other sharps injury), mucocutaneous injury (splashes of blood or other body fluids into the eyes, nose or mouth) or blood contact with damaged skin [1]. Healthcare Workers (HCWs) are at risk for occupational exposure to blood borne pathogens, including hepatitis B virus (HBV), hepatitis C virus (HCV), and human immunodeficiency virus (HIV). In 2012, RAISIN(Réseau d'Alerte, d'Investigation et de Surveillance des Infections Nosocomiales, in English: Alert, Investigation and Surveillance Network for Nosocomial Infections) found that the major risk from exposure to blood and body fluids is shown in descending order by HBV in 30%, HCV in 3% and HIV in 0.3% [2]. The risk of accidental contact with the blood and body fluids is especially increased in the following situations: while taking blood samples, during intravenous cannulation, intramuscular or subcutaneous injection, recapping of already used needle(s), surgery – especially during wound closure, and during clean up and transportation of waste materials [3].

In industrialized countries, accidents due by a sting are in the majority of contamination and accidental seroconversions with HBV and HCV principally involve nurses (47%) and laboratory workers (22%) [4]. In particular, Badidi *et al.* have carried out a survey among the health professionals in two hospital structures of Lubumbashi to evaluate their knowledge and practices in front of the Exposure Accidents to Blood (during May 2013). The study reported that 81.3% of health professionals had a good knowledge of the risk of contamination three HBV, HCV, and HIV. Furthermore, authors found that accidents rate of blood exposure were 21.6%. 61.2% during vaccination against hepatitis B. The accidents happened during recapping of needle. The authors concluded that it was necessary to increase the education and awareness of health professionals to reduce these kinds of incidents [4]. However, a study conducted in West Africa estimated the incidence of BEA at1.8/surgeon/year, 0.6/nurse/year and 0.3/physician/year [5].

A study in India showed the highest incidence of occupational exposure among nurses [6]. It has been reported that nurses experience the majority of needle-stick injuries in the world including half of the exposures that occur in the US [7, 8] and 70% of exposures occurring in Canada [9].

Other studies in Algeria showed that the number of BEA is still unknown, due in particular to the absence of a continuous monitoring system for this type of accident in our healthcare facilities and even to the under-reporting of the accident by HCWs. Only a study conducted out over two years (January 2005 – December 2006) shows that the incidence of BEA was 44 accidents reported in 2005 and 64 in 2006. Needlestick injuries represented 81% of cases [10]. These infections are preventable through infection control measures, which significantly reduce the risk of HIV and Hepatitis transmission among health workers [11].

In the hospital of Batna, few actions have been carried out so far in the prevention of the accident of exposure to blood (safety equipment, personal protective equipment, sharps collectors...*etc.*). Therefore, the aim of this study was to evaluate the knowledge, attitudes and practices of staff working at the Emergency Department of the University Hospital of Batna city in front of the exposure accidents to blood. The Emergency Department is selected on the basis of the nature of the care provided and the large number of staff who are victims of BEA.

## DEFINITIONS OF HEALTHCARE PERSONNEL (HCP) AND BEA

HealthCare workers(HCWs) refers to all paid and unpaid persons working in healthcare settings who have the potential for exposure to infectious materials, including body substances (*e.g.*, blood, tissue, and specific body fluids), contaminated medical supplies and equipment, and contaminated environmental surfaces. HCWs might include but are not limited to emergency medical service personnel, dental personnel, laboratory personnel, autopsy personnel, nurses, nursing assistants, physicians, technicians, therapists, pharmacists, students and trainees, contractual staff not employed by the healthcare facility, and persons not directly involved in patient care but potentially exposed to blood and body fluids (*e.g.*, clerical, dietary, housekeeping, security, maintenance, and volunteer personnel) [12].

The term accident implies "exposure of a health care worker (HCW) to blood or body fluids through percutaneous lesions or through the introduction of the blood or a body fluid by way of the mucous membrane or skin lesions" [13]. Blood Exposure Accidents (BEA) are defined as the unintended contact with blood and or body fluids mixed with blood during a medical intervention. It carries the risk of infection by numerous blood-borne viruses [13, 14]. Exposures occur through needle sticks or cuts from other sharp instruments contaminated with an infected patient's blood or through contact of the eye, nose, mouth, or skin with a patient's blood [15].

# REGULATORY INSTRUMENTS RELATED TO BLOOD EXPOSURE ACCIDENTS (BEA) IN ALGERIA

- Act N°88-07 of January 1988 on health and safety and occupational medicine.
- Executive Decree N°91-05 of 19 January 1991 on the general protection requirements applicable to health and safety at work.
- Executive Decree N°93-120 of 15 May 1993 on the organization of occupational medicine.
- Act N°01-19 of 12 December 2001, on the management, control and disposal of waste.
- Executive Decree no. 02-427 of 07 December 2002, relating to the conditions for organizing instruction, information, and training
- Executive Decree no. 02-427 of 07 December 2002, concerning the conditions for organizing, the instruction, information and training of workers in the field of the prevention of occupational risks.
- Instruction No. 61 of 25 January 2000 on vaccination in the workplace. Order of 25 April 2000, concerning the vaccination against hepatitis B.
- Ministerial Instruction N°14 of 10 September 2002 on the obligation of vaccination against HBV.
- Ministerial Instruction 002 of 25 January 2004 on the health protection of students and staff in paramedical training schools.
- Ministerial Instruction No. 1355 MSPRH/DP/ of 06 June 2005, relating to the prevention of accidents involving exposure to blood (BEA) in healthcare settings.
- Ministerial Instruction N°002/MSPRH/MIN of 21 March 2006, relating to the prevention of transmission of hepatitis B and C virus in the care setting.

## SUBJECTS AND METHODS

Batna city located at 435 Kms in the South of the capital of Algeria. Its surface is 12.038,76 km$^2$ (according to land register of Batna). The Climate of Batna is one of a semi-arid region. The average temperature is of 4°C in January and of 35°C in July. The Hospital-University Centre of Batna, is a structure of about 612 Beds including 09 Medical Services, 06 Surgical Services, a Medical-Surgical Emergency Pavilion, and a technical tray of:08 laboratories, Radiology with 03 Appendices that provides: Scanner, I.R.M, Abdominal Echo, Echo-Doppler, Echo-Cervical, Echo-Mammary, Mammography, and six (06) Operating Rooms with 22 Operating Rooms.

Our study was conducted at the Emergency Department. Emergencies are the hospital service responsible for emergency care and it is a 72-bed structure organized: Medical Consultation, Surgical, Laboratory, Radiology, Deceleration and Preoperative Unit. The staff of Emergency Department is 289.

The work is a descriptive and epidemiological study that involved the health personnel of the emergency department to assess their knowledge of Blood Exposure Accidents. To do this, we used a questionnaire. Although some difficulties were encountered, namely, data collection, where some people refused to complete our questionnaire due to time constraints, or workload, others, mainly those working in the operating room, were not always accessible. But, with good perseverance, we had a significant population for our survey.

The target population is made up of all the health personnel working in the emergency department of Batna University Hospital: doctors, nurses, laboratory technicians, and maids. This is accidental sampling. The staff that completed the questionnaire was those who were present in the function department at the time of our survey.

The questionnaire consisted of 33 questions (open, semi-open, closed) that assessed health workers' knowledge of BEA. The data were collected through an anonymous individual questionnaire, previously evaluated, and validated during a pilot survey with a few professionals to assess the understanding of the questions and estimate the time for each survey. The questionnaire was distributed to health workers for completion in their functional departments after clearly outlining the objectives of the study.

Data collection took place over a period of 1 month, from the end of February 2019 to the end of March 2019. The number of health professionals responding was 160. Our investigation was based on the following (characteristics of the target population, compliance with good hygiene rules, identification of the BEA, assessment of knowledge about BEAs, assessment of attitudes, procedures to follow in the event of a BEA).


## RESULTS

### Socio-Professional Characteristics of Respondents

We started our study by the distribution of respondents by socio-professional characteristics (Table **1**).

**Table 1. Distribution of surveyed subjects according to the socio-professional characteristics.**

| Variable | Effective (n=160) | Percentage of data |
|---|---|---|
| Age | | |
| 20-30 | 87 | 54.37 |
| 31-40 | 33 | 20.63 |
| 41-50 | 18 | 11.25 |
| ≥ 50 | 22 | 13.75 |
| Total | 160 | 100 |
| Sex | | |
| Male | 70 | 43.75 |
| Female | 90 | 56.25 |
| Total | 160 | 100 |
| Seniority (year) | | |
| 1-5 | 86 | 53.75 |
| 6-10 | 30 | 18.75 |
| 11-15 | 10 | 6.25 |
| 20-25 | 4 | 2.5 |
| Plus de 26 | 30 | 18.75 |
| Total | 160 | 100 |
| Professional category | | |
| Nurses | 114 | 71.25 |
| Doctor | 30 | 18.75 |
| Laboratory technician | 10 | 6.25 |
| Maids | 6 | 3.75 |
| Total | 160 | 100 |

The female sex is predominant, with 56.25% of the cases. The most common age range is 20-30 years, with 54.37% of respondents. 53.75% of the respondents had seniority of between 1 and 5 years and 18.75% had seniority of more than 26 years. The distribution of the respondents according to the service shows that nurses represent the most affected occupational group with 71.25%, doctors with 18.75%, laboratory and maids are respectively 6.25% and 3.75% with the minority.

## History of Blood Exposure Accidents

Considering that we studied 160 Healthcare Workers (HCWs), 87.5% of respondents believe that accidents of blood exposure are frequent in the emergency department. 75% of the respondents were victims of at least one accident of blood exposure during their professional practice in the past 5 years due to the significant occurrence of this accident in their services and this by frequencies: 21.25% for a single exposure, 15% for 2 exposures, 6.25% for 3 exposures, and 32.5% for more than 4 exposures. The history of blood exposure accidents of our subjects is showing in Table **2**.

**Table 2. Distribution of subjects surveyed according to History of Blood Exposure Accidents.**

| Variable | Effective (n=160) | Percentage of Data |
|---|---|---|
| **Accidents of blood exposure are frequent in the emergency department?** | | |
| Yes | 140 | 87.5 |
| No | 20 | 12.5 |
| How many accidents of blood exposure have you been exposed to in the last 5 years? | | |
| 00 | 40 | 25 |
| 01 | 34 | 21.25 |
| 02 | 24 | 15 |
| 03 | 10 | 6.25 |
| 4 and more | 52 | 32.5 |
| Nature of the exposure accidents to blood | | |
| Needle sting | 136 | 85 |
| Skin cut by a sharp object | 18 | 11.25 |
| Projection of blood or biological fluid contaminated with blood | 6 | 3.75 |
| Circumstances of occurrence of exposure accidents to blood | | |
| Blood collection and injections (recapping needles) | 114 | 71.25 |
| In some very difficult care situations (transfusion, perfusion and dressing) | 18 | 11.25 |
| Waste handling and transport | 28 | 17.5 |
| The following circumstances predispose to an exposure accident to blood? | | |
| Duration of the intervention | 20 | 12.5 |
| Hemorrhagic characteristics of the operation | 58 | 36.25 |
| Night work in the emergency room | 45 | 28.12 |
| Patient's serological status | 37 | 23.13 |

For different exposures and for the seventy-five percent of the respondents who were victims of the accidents of blood exposure during their professional practice, 85% of the accidents of blood exposure were due to a needle bite, because needles are used in almost every act of care. The cutaneous cut by a sharp object: 11.25%. The projection of blood or biological fluid contaminated with blood: 3.75%. With 71.25% of the accidental contact with blood and body fluids are especially due to the following situations: while taking blood samples, during intravenous, intramuscular or subcutaneous injection, recapping of already used needle(s), surgery – especially during wound closure, and with 17.5% during the clean up and transportation of waste materials. On the other hand, some circumstances predispose to an exposure accident to blood. As well as having quick thinking and clinical skills, medical staffs are often extremely busy, and have to be calm when dealing with angry people who might have been waiting a long time. Also, other circumstances as the hemorrhagic characteristics of the operation with 36.25% of our respondents, the night work in the emergency room: 36.25% and the patient's serological status: 23.13%.

## Knowledge Assessment of Blood-Borne Infectious Agents

Regarding the risks of transmission of pathogens after an accident of blood exposure, we noted that HCV, HIV, HBV were the most well known by our population (Table **3**) with respective percentages of 61.25%, 25%, 13.75%, due to their significant spread between patients. The transmission of infections by blood was known. Then, during an accident of blood exposure, 59.38% find that the severity of infection in the source patient was the most factors to assess the risk of transmission, our interviewees insist on the seriousness of the infection in the source patient because they are afraid of being infected. While 26.87% believe that the type of equipment involved (sharp or not) is the factor of the risk.

Of the 160 respondents, 80% felt that reporting is not done in the case of exposure accidents to blood, and according to 42.51% of the respondents who non-reporting, they didn't do it because they didn't find the time since the declaration must be made within 48 hours to the administration as an industrial accident. However, 29.37% of them find the procedure too complex.

**Table 3. Distribution of subjects surveyed according to knowledge about blood-borne infectious agents and post-exposure.**

| Variable | Effective (n=160) | Percentage of Data |
|---|---|---|
| What are the most dangerous infections? | | |
| HCV | 98 | 61.25 |
| HBV | 22 | 13.75 |
| HIV | 40 | 25 |
| Knowledge of risk of transmitted blood-borne infections | | |
| Yes | 160 | 100 |
| No | 0 | 0 |
| On what do you assess the risk of transmission during an accident of blood exposure? | | |
| On the depth of the wound | 22 | 13.75 |
| Type of equipment involved (sharp or not) | 43 | 26.87 |
| Severity of infection in the source patient | 95 | 59.38 |
| Declaration of exposure accidents to blood | | |
| Yes | 32 | 20 |
| No | 128 | 80 |
| Reasons for non-reporting | | |
| You didn't find the time | 68 | 42.51 |
| You feel there is no risk | 31 | 19.37 |
| You are afraid of the sanction, of the judgment of others | 14 | 8.75 |
| You find the procedure too complex | 47 | 29.37 |

## Action to be Taken after Being the Victim of an Accident of Blood Exposure

Ninety decimal point eighty-four percent of respondents who have been victims of an accident of blood exposure say that our reaction following the accident is stopping their activities and perform first aid to ensure their safety, while 14.16% do not react immediately because they think that there is no risk. And, so, most of them had disinfected the wound, 36.66% with alcohol and 30.84% use Betadine. Then, 20.84% who used bleach and 11.66, who had done a simple wash with soap. According to the respondents, the majority of staff (75%) responded that knowledge of the patient's positive serological status changes their protective behaviors because the risk of contamination is higher. However, only 15% of our population have benefited from serological follow-up after an accident of blood exposure, while 85% have not benefited from serological follow-up and this is due to the lack of information and awareness. Related to the existence of a protocol to be followed in the event of an accident of blood exposure, 80% of respondents stated that there is no protocol to follow after an ABE in their department and this is due to the negligence of managers in terms of training and awareness. Also, the majority of the respondents (85.84%) say there's no referring doctor.

**Table 4. Distribution of subjects surveyed according to the action to be taken after being victims of an accident of blood exposure.**

| Variable | Effective (n=120) | Percentage of Data |
|---|---|---|
| What was your reaction following the accidents of blood exposure | | |
| No immediate reaction | 11 | 9.16 |
| Stopping the activity and performing first aid | 109 | 90.84 |
| The most appropriate immediate care after an accident of blood exposure | | |
| Use a bleach | 25 | 20.84 |
| Disinfection with Betadine | 37 | 30.84 |
| Disinfection with alcohol | 44 | 36.66 |
| Simple washing with soap | 14 | 11.66 |
| Search for the patient's serological status | | |
| Yes | 90 | 75 |
| No | 30 | 25 |
| HIV, HBV, HCV serological follow-up after an accidents of blood exposure | | |
| Yes | 18 | 15 |
| No | 102 | 85 |
| Existence of a protocol to be followed in case of an accidents of blood exposure | | |
| Yes | 24 | 20 |
| No | 96 | 80 |
| Existence of a referring doctor to take care of an accidents of blood exposure victims | | |
| Yes | 17 | 14.16 |
| No | 103 | 85.84 |

## Measures for Prevention of Blood Exposure Accidents

One hundred percent of our respondents used soap when washing hands, and the majority of them (95%) answered that they wear gloves for care to ensure their protection, but (5%) answered that they do not wear them due to negligence. After care, only 41.25% of respondents do not recapping needles after use because they are afraid of accidents secondary to needle stick injuries when replacing needle caps. While 58.75% of them recap the needles to be safe when handling waste.

For the availability of sharps collectors in the workplace, 96.25% of respondents confirm the existence of sharps collectors in the workplace, while others (3.75%) say that they don't exist or their volume is not adequate (large volume collectors difficult to mobilize).

The treatment of biomedical waste, one hundred percent confirmed incineration as a treatment method.

Concerning vaccination, most respondents (87.5%) were vaccinated against hepatitis B compared with 12.5% who were unvaccinated. But the efficiency of vaccination (2-3 doses) was done just for 70% of them (Table **5**).

**Table 5. Distribution of subjects surveyed according to the prevention of Blood Exposure Accidents (BEA).**

| Variable | Effective (n=160) | Percentage of data |
|---|---|---|
| Use of soap during washing | | |
| Yes | 160 | 100 |
| No | 0 | 0 |
| Wearing gloves for the care | | |
| Yes | 152 | 95 |
| No | 8 | 5 |
| Recapping the needles after use | | |
| Yes | 94 | 58.75 |
| No | 66 | 41.25 |
| Availability of sharps collectors in the workplace | | |
| Yes | 154 | 96.25 |
| No | 6 | 3.75 |
| Elimination of Sharp Objects in their collectors | | |
| Yes | 127 | 79.37 |
| No | 33 | 20.63 |
| Treatment of biomedical waste | | |
| Incineration | 160 | 100 |
| Burying | 0 | 0 |
| Hepatitis B Vaccine Status | | |
| Vaccinated | 140 | 87.5 |
| Not vaccinated | 20 | 12.5 |

*(Table 5) cont.....*

| Variable | Effective (n=160) | Percentage of data |
|:---:|:---:|:---:|
| If yes, is your vaccination complete (3-4 doses)? | | |
| Yes | 98 | 70 |
| No | 42 | 30 |

## DISCUSSION

Analysis of various responses obtained from the staff of the emergency department of the university hospital of Batna city showed that:

Our sample is represented by a relatively young population with professional seniority of less than 05 years, which may explain the lack of experience, therefore, greater risk of an accident of blood exposure.

87.5% of respondents believe that blood exposure accidents are frequent in the emergency department because of the large number of patients requiring care, the speed of the gestures and the frequency of use of prickly or sharp equipment. According to our results, bites are the majority, as found in various studies [16].

Nurses had a higher number of accidents during needle recapping and handling of contaminated sharp devices. They had significantly more frequent contact with blood through damaged skin, while doctors had more contact with blood through the conjunctiva or mucous membranes. As well as having the highest prevalence of percutaneous and mucocutaneous injuries among HCWs, doctors and nurses also have the highest rates of underreporting of these injuries among HCWs [17 - 19].

The most commonly reported reasons for non-reporting in the literature are perceived low risk of infection and lack of time [20, 21].

Most of the people questioned (78.75%) think that the level of compliance with hygiene measures within their department is insufficient, and this may come down to lack of equipment and means, equipment of poor quality or negligence, or workload. 75% of respondents were victims of an accident of blood exposure at least once, of which 32.5% were exposed to more than 4 accident of blood exposure. A number of factors were be associated with increased exposure to BEA, they include working in the emergency department, working at duty posts where more clinical procedures are done, working night.

The majority of staff (95%) indicated that they wear soiling gloves to protect themselves, and (5%) indicated that they do not wear them because of neglect or lack of gloves. The majority of staff (87.5%) indicated that they wear only gloves as a means of personal protection because they are the only means available. The increasing severity of blood exposure accidents is linked to the lack of safe behavior against this risk. It is obvious that then, only the widest possible implementation of high-performance safety equipment will be able to provide a strong and sustainable response in terms of controlling the risk of blood exposure accidents [22].

Employers should have in place a system that includes written protocols for prompt reporting, evaluation, counseling, treatment, and follow-up of occupational exposures that may place a worker at risk of blood-borne pathogen infection [23].

The majority of staff (83.75%) indicated that knowing the patient's positive HIV status changes their protective behaviors because the risk of contamination is higher. Regarding the risk assessment of transmissions during an accident of blood exposure, our interviewees stress the severity of the infection in the source patient: 72.5% because they are afraid of being infected. 88% are vaccinated against hepatitis B, which is a positive element that will help prevent complications related to an accident in an HBV-positive patient. According to 73% of staff, there is neither a protocol to follow in the event of an accident of exposure to Blood, nor a doctor referral in their departments; this demonstrates the absence of a conscience on the part of the whole team of the importance of protocols.

As a final point, we suggest this strategy to protect HCWs from occupational exposures to blood and body fluids in the event of a blood exposure accident:

- Stop the activity and do the immediate care: immediate wound cleaning and disinfection,
- Source patient (HIV) screening,
- Assessment of other infectious risks: HBV and HCV,
- Declaration of work accident within 24 to 48 hours,
- Biological surveillance (assessment: NF, creatinemia, transaminases, HIV serology, HCV, *etc.*).

## CONCLUSION

In hospitals, and in the service of patients, health workers sometimes forget that they themselves are potential patients and, by "custom", often impose working and living conditions that are contrary to what they can advise the patients they are treating. Exposure to blood-borne pathogens (BBPs) continues to pose a significant risk to healthcare workers (HCWs) [24], despite the introduction of regulatory strategies designed to decrease the incidence and danger from such exposure [25] and the use of safer devices (*i.e.* engineering controls) [26]. These blood-borne infections have serious consequences, including long-term illness, disability, and death [27, 28]. Thus, in 1987, CDS (Centers for Disease Control) developed universal precautions to help protect both HCWs and patients from infection with blood-borne pathogens in the health care setting [29].

The study showed that there is experiencing real suffering of the accident of blood exposure in the Emergency Department and especially their nature, namely hepatitis. Medical staffs are trained to work quickly and effectively, even with minimal information. The most common accidents were needle stick injuries, and those resulting from improper handling of contaminated sharp devices [30]. Consequently, we insist on continuous and targeted education of health care workers on the risks of percutaneous and mucocutaneous injuries, including the acquisition of blood borne pathogens [31]. Furthermore, standard precautions could be supplemented by educating health-care workers to take responsibility for their own health and safety and for that of others who may be affected by their actions at work [32].

Although the government has promulgated BEA law, the application of this law remains limited; Finally, and due to the importance and frequency of accident of blood exposure and the potential infectious risk, and their clear under-reporting among health personnel (most commonly among nurses) at the emergency department, it is necessary and indispensable to set up a reporting circuit for these accidents. Reporting injuries and documenting all blood-borne exposures are essential for having the evidence to analyze for prevention. Therefore, a study extended to all the departments of the University hospital of Batna seems necessary to evaluate the frequency of BEA, their management, and the technical and medical preventive measures to reduce them.

## CONSENT FOR PUBLICATION

Not applicable.

## CONFLICT OF INTEREST

The authors confirm that the contents of this chapter have no conflict of interest.

## ACKNOWLEDGEMENTS

The authors hereby thank the authorities and staff of the Emergency department of the University Hospital of Batna and all who participated in the study.

# REFERENCES

[1]     World Health Organization. Health care worker safety: AIDE-MEMOIRE for a strategy to protect health workers from infection with bloodborne viruses 2013.http://www.who.int/injection_safety/toolbox/en/AM HCW_Safety_EN.pdf

[2]     Blood Exposure Study Group. Task Force on Blood Exposure. National Surveillance of Blood Exposure Accidents-RAISIN 2012.

[3]     Markovic-Denic L, Maksimovic N, Marusic V, Vucicevic J, Ostric I, Djuric D. Occupational exposure to blood and body fluids among health-care workers in Serbia. Med Princ Pract 2015; 24(1): 36-41. [http://dx.doi.org/10.1159/000368234] [PMID: 25376432]

[4]     Babidi BL, Bakadia BM, Kalenga MP, *et al.* Evaluation of knowledge, attitudes and practices of health professionals in front of the exposure accidents to blood in two hospital structures of lubumbashi. Open Access Library Journal 2017; 4: e3823. [http://dx.doi.org/10.4236/oalib.1103823]

[5]     Tarantola A, Koumaré A, Rachline A, *et al.* Groupe d'Etude des Risques d'Exposition des Soignants aux agents infectieux (GERES). A descriptive, retrospective study of 567 accidental blood exposures in healthcare workers in three West African countries. J Hosp Infect 2005; 60(3): 276-82. [http://dx.doi.org/10.1016/j.jhin.2004.11.025] [PMID: 16021690]

[6]     Singru SA, Banerjee A. Occupational exposure to blood and body fluids among health care workers in a teaching hospital in mumbai, India. Ind J Comm Med 2008; 33(1): 26-30. [http://dx.doi.org/10.4103/0970-0218.39239] [PMID: 19966992]

[7]     Centers for Disease Control and Prevention. Workbook for designing, implementing and evaluating a sharps injury prevention program 2004.http://www.cdc.gov/sharpssafety/pdf/WorkbookComplete.pdf

[8]     Prüss-Üstün A, Rapiti E, Hutin Y. Sharps injuries: Global burden of disease from sharps injuries to health-care workers 2003.http://wwwwhoint/peh/burden/9241562463/sharptochtm

[9]     Canadian Center for Occupational Health and Safety (CCOHS) Needlestick injuries 2000.http://www.ccohs.ca/oshanswers/diseases/needlestick_injuries.html

[10]    Beghdadli B, Ghomari O, Taleb M, *et al.* Le personnel a risque d'accidents d'exposition au sang dans un CHU de l'Ouest Algérien. Sante Publique 2009; 21(3): 253-61.https://www.cairn.info/revue-sant--publique-2009-3-page-253.htm [http://dx.doi.org/10.3917/spub.093.0253] [PMID: 19863016]

[11]    Gupta A, Anand S, Sastry J, *et al.* High risk for occupational exposure to HIV and utilization of post-exposure prophylaxis in a teaching hospital in Pune, India. BMC Infect Dis 2008; 8(8): 142. [http://dx.doi.org/10.1186/1471-2334-8-142] [PMID: 18939992]

[12]    David T, Kuhar , David K, *et al.* US Public Health Service Working Group. Updated US Public Health Service Guidelines for the Management of Occupational Exposures to Human Immunodeficiency Virus and Recommendations for Postexposure Prophylaxis 2013; 34(9): 875-92.

[13]    Siegel JD, Rhinehart E, Jackson M, Chiarello L. The healthcare infection control practices advisory committee. Guideline for Isolation Precautions: Preventing Transmission of Infectious Agents in Healthcare Settings https://www.cdc.gov/infectioncontrol/guidelines/isolation/index.html

[14]    International Labour Organization (ILO). An ILO code of practice on HIV/AIDS and world of work. Geneva: ILO 2001; p. 38.

[15]    Kunkel D. Exposure to Blood, What Healthcare Personnel Need to Know, Department of Health & Human Services, Centers for Disease Control and Prevention. 2003.

[16]    RAISIN, GERES. Surveillance des accidents avec exposition au sang. Guide méthodologique. Réseau d'Alerte, d'Investigation et de Surveillance des Infections Nosocomiales RAISIN et GERES 2004; 43 pages.http://www.invs.sante.fr/raisin/

[17]    Lachowicz R, Matthews PA. The pattern of sharps injury to healthcare workers at Witbank hospital. S Afr Fam Pract 2009; 51(2): 148-51.
[http://dx.doi.org/10.1080/20786204.2009.10873831]

[18]    Voide C, Darling KE, Kenfak-Foguena A, Erard V, Cavassini M, Lazor-Blanchet C. Underreporting of needlestick and sharps injuries among healthcare workers in a Swiss University Hospital. Swiss Med Wkly 2012; 142: w13523.
[http://dx.doi.org/10.4414/smw.2012.13523] [PMID: 22328010]

[19]    Evans B, Duggan W, Baker J, Ramsay M, Abiteboul D. Exposure of healthcare workers in England, Wales, and Northern Ireland to bloodborne viruses between July 1997 and June 2000: analysis of surveillance data. BMJ 2001; 322(7283): 397-8.
[http://dx.doi.org/10.1136/bmj.322.7283.397] [PMID: 11179157]

[20]    Burke S, Madan I. Contamination incidents among doctors and midwives: reasons for non-reporting and knowledge of risks. Occup Med (Lond) 1997; 47(6): 357-60.
[http://dx.doi.org/10.1093/occmed/47.6.357] [PMID: 9327639]

[21]    Haiduven DJ, Simpkins SM, Phillips ES, Stevens DA. A survey of percutaneous/mucocutaneous injury reporting in a public teaching hospital. J Hosp Infect 1999; 41(2): 151-4.
[http://dx.doi.org/10.1016/S0195-6701(99)90053-1] [PMID: 10063478]

[22]    Lamontagne F, Lolom I, Tarantola A, Descamps JM, Bouvet E, Abiteboul D. Evolution de l'incidence des accidents exposant au sang chez le personnel infirmier hospitalier en France métropolitaine de 1990 à 2000: impact des mesures préventives et des matériels sécurisés. Hygiènes 2003; XI: 113-9.

[23]    Beltrami EM, Williams IT, Shapiro CN, Chamberland ME. Risk and management of blood-borne infections in health care workers. Clin Microbiol Rev 2000; 13(3): 385-407.
[http://dx.doi.org/10.1128/CMR.13.3.385] [PMID: 10885983]

[24]    Holodnick CL, Barkauskas V. Reducing percutaneous injuries in the OR by educational methods. AORN J 2000; 72(3): 461-464, 468-472, 475-476.
[http://dx.doi.org/10.1016/S0001-2092(06)61278-7] [PMID: 11004962]

[25]    Twitchell KT. Bloodborne pathogens. What you need to know--Part I. AAOHN J 2003; 51(1): 38-45.
[http://dx.doi.org/10.1177/216507990305100109] [PMID: 12596344]

[26]    Rogues AM, Verdun-Esquer C, Buisson-Valles I, *et al.* Impact of safety devices for preventing percutaneous injuries related to phlebotomy procedures in health care workers. Am J Infect Control 2004; 32(8): 441-4.
[http://dx.doi.org/10.1016/j.ajic.2004.07.006] [PMID: 15573049]

[27]    Rapiti E, Prüss-Üstün A, Hutin Y. Sharps injuries: assessing the burden of disease from sharps injuries to health-care workers at national and local levels WHO Environmental Burden of Disease Series, No 11. Geneva: World Health Organization 2005.

[28]    Prüss-Ustün A, Rapiti E, Hutin Y. Estimation of the global burden of disease attributable to contaminated sharps injuries among health-care workers. Am J Ind Med 2005; 48(6): 482-90.
[http://dx.doi.org/10.1002/ajim.20230] [PMID: 16299710]

[29]    Centers for Disease Control (CDC). Recommendations for prevention of HIV transmission in health-care settings. MMWR Morb Mortal Wkly Rep 1987; 36(2): 1S-18S.
[PMID: 3112554]

[30]    Marković-Denić L, Branković M, Maksimović N, *et al.* Occupational exposures to blood and body fluids among health care workers at university hospitals. Srp Arh Celok Lek 2013; 141(11-12): 789-93.
[http://dx.doi.org/10.2298/SARH1312789M] [PMID: 24502099]

[31]    Mbah CC. Reporting of Accidental Occupational Exposures to Blood and Body Fluids by Doctors and Nurses in the Public Primary Health Care setting of sub district F of Johannesburg metropolitan district, A research report submitted to the Faculty of Health Sciences, University of the

Witwatersrand. Johannesburg, South Africa: Johannesburg, Master of Medicine in Family Medicine 2014.http://wiredspace.wits.ac.za/handle/10539/15293

[32] Auta A, Adewuyi EO, Tor-Anyiin A, *et al.* Health-care workers' occupational exposures to body fluids in 21 countries in Africa: systematic review and meta-analysis. Bull World Health Organ 2017; 95(12): 831-841F.
[http://dx.doi.org/10.2471/BLT.17.195735] [PMID: 29200524]

**CHAPTER 4**

# Hazardous Waste Management in India: Risks and Challenges Associated with Hazardous Waste

**Arvind Kumar Shukla**[1,2,3,*] and **Sandhya Shukla**[3]

[1] *School of Biotechnology and Bioinformatics, D.Y. Patil University, Plot No.50, Sector- 15, C.B.D. Belapur, Navi Mumbai, 400614, Maharashtra, India*

[2] *School of Biomedical Convergence Engineering, Pusan National University, Yangsan 50612, Korea*

[3] *Inventra Medclin Biomedical Healthcare and Research Center, Katemanivli, Kalyan, Thane, 421306, Maharashtra, India*

**Abstract:** Today, there are major environmental challenges associated with hazardous waste management in India. In this regard, hazardous production is increasing rapidly and it's required proper management to be improving the future of India. In this context, the Indian government's policy and legislation are not working properly on Hazardous Waste Management (HWM), thereby triggering enormous public health issues. The effective Hazardous waste (HW) treatment practices are of paramount importance to protect human health, and in addition, this is a social responsibility for the safety of the future generation. This chapter reviews the current situation on the rules and regulations of HWM and policy to adopt new successful approaches and techniques that are used worldwide to address this issue in India. Current systems in India cannot cope with the volumes of waste generated by an increasing urban population because of the lack of adequate infrastructure but these challenges will allow to transforming barriers in good opportunities for sustainable growth.

**Keywords:** Biomedical hazardous waste, Hazardous waste management, India, Municipal hazardous solid waste, Urbanization.

## INTRODUCTION

Many developing countries such as India are facing many difficulties in the management of hazardous waste with related health problems such as cancer, diabetes and other infectious diseases. Currently, in India, continuous industrialization happens, therefore, a higher amount of hazardous waste is pro-

* **Corresponding author Arvind Kumar Shukla:** School of Biotechnology and Bioinformatics, D.Y. Patil University, Plot No.50, Sector- 15, C.B.D. Belapur, Navi Mumbai, 400614, Maharashtra, India; E-mail: arvindkumarshukla10@gmail.com

duced because of poor management systems and lack of adoption of technologies to overcome this problem.

The first example of this situation is an accident at India's Bhopal plant, methyl isocyanate (MIC) gas leak accident, which is the responsible of the Union Carbide Corporation of the USA. The deaths of at least 2,000 citizens resulted from operational accidents, structural defects, repair deficiencies, training shortages, and safety-threatening economic steps, according to current and former staff, business scientific records, and the Chief Scientist of the Indian Government [1].

Currently, in India, manufacturing operations, farming and agro-factories, medical facilities, shopping centers, households, and the informal sector are the primary sources of hazardous solid waste, but they are neither recognized nor regulated by the Indian government. Hazardous waste has been described as "any material, including domestic and radioactive waste, which, due to its quantity and/or corrosive, reactive, ignitable, poisonous and contagious characteristics, can cause harmful environmental and human health consequences. This material is non-degradable, highly toxic, and even lethal at very low concentrations. It can be biologically magnified and poses potential risks to human health [2].

Approximately 90% of residual waste is currently dumped instead of properly landfilled [3, 4]. In this case, the more successful proactive HWM must be followed immediately. To do this, a new management structure and hazardous waste disposal systems had to be implemented. These days, the HWM program is an inefficient and hazardous waste has a negative impact on public health, the atmosphere, and the economy [5]. India's Ministry of Environment and Forests (MOEF) is enforcing Hazardous Waste Management laws and managing rules [6], despite the fact that enforcement is variable and restricted. This chapter introduces the new obstacle, challenges, and possibilities associated with the development of a new scenario for hazardous waste management in India.

## HAZARDOUS WASTE MANAGEMENT IN INDIA

At this moment, the lack of infrastructure and the limitation of governmental and non-governmental bodies that regulate and manage hazardous waste in India's various cities has resulted in ineffective management [7, 8, 16]. Waste contractors collecting hazardous waste from various industries are mostly poorly equipped to handle radiated, carcinogenic chemicals, heavy metals, industrial oils, lubricants, grease, untrained and poorly paid, and India's high-temperature treatment infrastructure is inadequate [9, 10, 16]. Burning of agricultural biomass residue or crop residue burning (CRB), landfill burning is still very common disposal practices contributing to damage to human health and the environment. Currently, India's urban states such as Maharashtra, Gujarat, Andhra Pradesh, Tamil Nadu face problems with rising levels of hazardous waste [11 - 13, 16]. Gujarat state contributes to one of India's fastest-growing industrial development areas, with high quantities of chemicals, petrochemicals, pharmaceuticals, drugs, pesticides, fertilizers, textiles, and paper factories. Untreated waste from these industries is generally the major cause of the state's pollution. Therefore, the Gujarat state is one of India's highest producers of hazardous waste. Untreated waste from these industries is generally the primary cause of the state's pollution.

Recent data indicates that India generates about 51.1 MMT of waste annually, with about 7.46 MMT of hazardous waste generated from 43,936 industries [12, 16]. About 0.69 MMT (9%) is incinerated, 3.41 MMT (46%) is landfilled and 3.35 MMT (45%) is recycled. Gujarat is one of India's states that produces the highest quantity of hazardous waste such as 7751 hazardous waste producing small and large industries contributing to about 28.76% of India's waste generation. More than 70% of all industrial hazardous waste from thermal power plants, which is generated by coal ash [14 - 16].

Ineffective management of solid waste is a major problem in India, particularly in urban centers [64, 65]. Through 2050, around 50 percent of India's population is expected to live in urban areas, and waste production is expected to grow through 5 percent every year [66, 67]. About 101 million metric tons (MMT), 164 MMT and 436 MMT per year are waste projected to appear by 2021, 2031 and 2050, respectively. The rise in the production of solid waste is primarily due to economic development, population growth and changing lifestyles (Fig. **1**).

Municipal solid waste, which is toxic and more hazardous, is often easily collected, transported, stored, and discarded or dumped without treatment or treatment. As a result, a considerable amount of waste remains unattended at recycling centers, roadsides and river banks, with many small and large manufacturing units disposing of their waste mostly in open areas and adjacent to water sources, resulting in environmental pollution and threats to public health [65]. Therefore, despite significant socio-economic advances, solid hazardous waste management systems have remained relatively unchanged and are inefficient in India. 3Rs (reduce, reuse and recycle) are scarcely used, despite being part of the country's policy framework. Several new legislation have been introduced by the Ministry of Environment and Forestry (MoEF & CC) to address some of these issues.

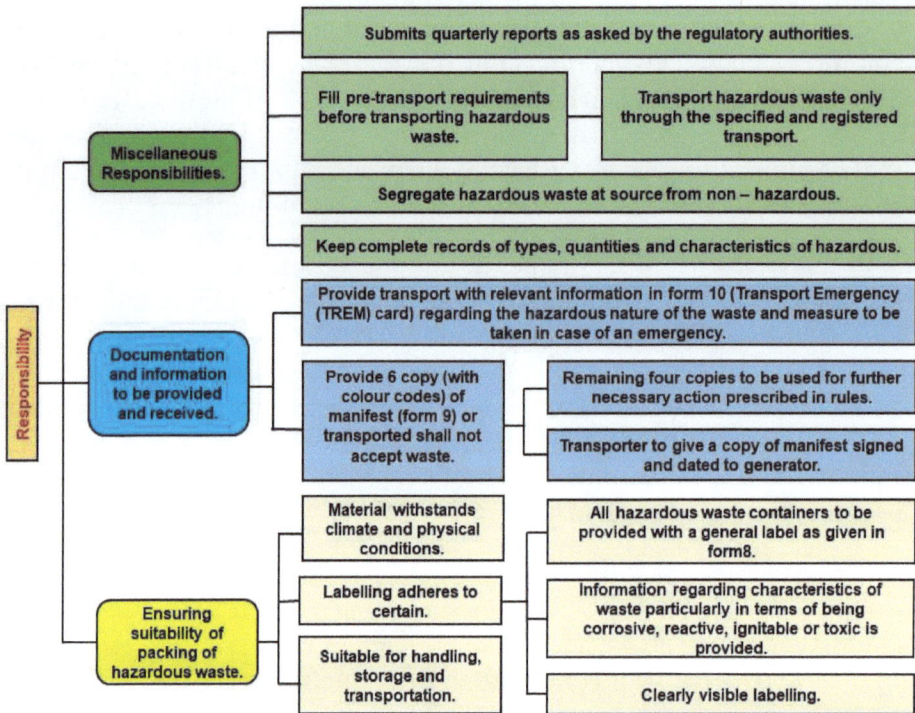

**Fig. (1).** The responsibilities of the generator of the hazardous waste. Source: ELI (2014) [9].

# POLICY AND REGULATIONS ON HAZARDOUS WASTE MANAGEMENT IN INDIA

Regulation and resource management of hazardous waste in India such as the Ministry of Environment and Forests (MoEF & CC) and the pollution control boards: the Central Pollution Control Board (CPCB) and the State Pollution Control Boards (SPCBs) together form the legislative and administrative heart of the Indian waste management sector [15, 16] (Table **1**). At this level, only Urban Local bodies are responsible for hazardous waste management. Regulation and permission from the respective SPCB are required for the management of industrial production of hazardous waste. The specific issue in the management of hazardous waste is that municipal authorities do not have the budgets to adequately cover the financial needs of efficient management systems. The current situation is that there is a scarcity of strategic plans and a shortage of governance structures such as separation, waste collection and control with significant obstacles to effective hazardous waste management (HWM) in India [9].

The Indian Constitution enacted in 1976, 42 amendment, which actually came and governed January 1977 and consequently directed the government to safeguard public health in order to conserve and improve the environment, forests and wildlife (Fig. **2**). The Constitution requires Directive Principles of State Policy (Article 47) not only to protect the state climate but also to enhance contaminated conditions effectively. In 2016, the toxic waste was transported from countries like Malaysia and Saudi Arabia [16].

**Table 1. The responsibilities of agencies for the Hazardous Waste Rules in India.**

| | Activity | Authority | | | |
|---|---|---|---|---|---|
| | | **MoEF & CC** | **SG** | **SPCB** | **CPCB** |
| 1 | Survey and inventorisation of hazardous waste generators and processors | | | X | |
| 2 | Grant authorisation for handling hazardous waste to sites and operators | | | X | |
| 3 | Inspect facilities/infrastructure/technical capabilities in hazardous waste units | | | X | |
| 4 | Suspend/refuse/can authorisation for handling hazardous waste | | | X | |
| 5 | Identify and notify sites for hazardous waste treatment/disposal facilities | | X | X | |
| 6 | Facilitate environmental impact assessment studies before identifying sites | | X | X | |
| 7 | Collect, collate and publish list of abandoned hazardous waste dump sites | | X | | |
| 8 | Establish a system for filing annual returns and reporting accidents by hazardous waste facilities and operators | | | X | |
| 9 | Process and grant permits for the import of hazardous waste to sites in India | | | X | |
| 10 | Examine and permit/refuse exporters' requests for the importation of hazardous waste into India | X | | X | |
| 11 | Issue instructions to hazardous waste importers | | | X | |
| 12 | Inform port authorities to take appropriate steps for the safe handling of hazardous waste at ports | X | | X | |
| 13 | Inspect records of imports | X | | X | X |
| 14 | Process appeals | | X | | |

**Source:** CPCB (Central Pollution Control Board); MoEF & CC (Ministry of Environment, Forests and Climate Change); SG (State Government); and SPCB (State Pollution Control Board). Adapted from [**15**].

In addition, the discontinued practice of hazardous waste management with other related waste types such as Management and Transboundary Movement Rules 2016 prohibiting the import of household waste, solid plastic waste, animal oils and edible fats. The land reservation for the disposal of sheds of hazardous waste includes the rules and regulations of the Indian government to ensure that staff engaging in the processing of hazardous waste are properly registered, provided with tools, learned expertise and paid, and that inspection authorities are formed to track the production and recycling of hazardous waste from each country [16].

The main piece of legislation is 1986's Environment (Protection) Act, which is a very powerful and essential regulation for protecting and improving the environment and controlling the treatment and storage of hazardous substances and chemicals. The Act also contains a number of laws, including Hazardous Waste (Management and Handling) Rules, 2008 (these are the primary regulations governing the management of hazardous waste in India), Biomedical Waste (Management and Handling) Rules 1998, Manufacture, Storage and Import of Hazardous Chemical Rules 2001 and E-Waste (Management and Handling) Rules 2010 [16, 17].

But presently, there are other major challenges such as staff shortages and financial resources, a lack of jurisdiction, and a lack of structured protocols [9, 16]. The Management and Trans-boundary Transport, Rules 2016 State of Dangerous and Other Wastes disposed of by property owners are liable to pay financial penalties if the rules for handling, disposal and reuse of such waste are not complied with and may even be jailed for misconduct. India's state governments have ordered specific rules to determine sites of constructing facilities for hazardous waste disposal. Moreover, no new sites are currently being built since the new rules entered into force. Many states in India, such as Punjab, Karnataka, Orissa, and Kerala, have no facilities for hazardous waste treatment [10, 16].

## CLASSIFICATION AND CHARACTERISTICS OF HAZARDOUS WASTES

Hazardous Wastes are classified as F, K, P, and U-Lists (Fig. **4**).

**F-List:** The F-list contains hazardous waste from non-specific sources, which is different industrial processes that could have produced the waste. The list includes solvents commonly used in degreasing, baths and sludges for metal treatment, wastewater from metal plating operations, and dioxin containing chemicals or their precursors. Examples: Benzene (F005), Cresylic acid (F004), Carbon tetrachloride (F001), *etc.*

**K-List:** The K-list contains hazardous waste generated through specific processes of industry. Examples include the manufacturing of wood, paint, chemical processing, petroleum refining, iron and steel production and pesticide production, amongst the industries that produce K-listed waste.

**P and U-Lists:** The P and U-Lists contain commercial chemical products discarded, chemicals off-specification, container residues, and material spillage residues. Those two categories include industrial pure chemical degrees, all chemical degrees produced or sold in technical terms, and all formulations that are the sole active ingredient of a product. A pesticide, which is not used during its shelf-life and needs mass storage, is an example of a P or U listed hazardous waste.

Furthermore, such waste has different characteristics such as:

## a). Ignitability

A material is a dangerous ignitable material when it has a flashpoint of less than 60 °C; readily catches the flame and burns to create a threat or it is a compressed gas or oxidizer that is inflammable. For example: naphtha, thinner lacquer, epoxy resin, adhesive, oil paint, *etc.*

## b). Corrosivity

Liquid waste with a pH of less than or equal to 2 or more than or equal to 12.5 is considered a corrosive hazardous wastage. In many industries, the cleaning and deterioration of metal components are often achieved by sodium hydroxide, a caustic solution with a high pH. Many industries use hydrochloric acid, a low pH solution, to purify metal parts before painting. If these caustic or acidic solutions have been eliminated, waste is a corrosive hazardous waste.

## c). Reactivity

A material is considered reactive hazardous waste when it is unstable, reacts violently to water, produces toxic gasses in the event of exposures to water, corrosion or if exposure to heat or flames could result in detonation or explosion. For example, sources of reactive waste are gunpowder, sodium metal and cyanide- or sulfide-containing waste.

## d). Toxicity

A representative sample of the material must be tested in a certified laboratory to determine if a waste is a toxic hazardous waste. The toxic characteristic identifies waste which may leach toxic substances into groundwater at a dangerous concentration (Table **2** Fig. **2**).

## Different class hazardous materials

Chemicals also require management as hazardous waste if they exhibit one or more of the four EPA-defined characteristics of **ignitability, corrosivity, reactivity, or toxicity**

## EPA Toxicity Characteristic List

| Heavy Metals | Pesticides | Common Organic Chemicals | |
| --- | --- | --- | --- |
| • Arsenic (5 ppm) | • Endrin (0.02 ppm) | • Benzene (0.5 ppm) | • Hexachlorobutadiene (0.5 ppm) |
| • Barium (100 ppm) | • Lindane (0.4 ppm) | • Carbon Tetrachloride (0.5 ppm) | • Hexachloroethane (3 ppm) |
| • Cadmium (1 ppm) | • Methoxychlor (10 ppm) | • Chlorobenzene (100 ppm) | • Methyl Ethyl Ketone (200 ppm) |
| • Chromium (5 ppm) | • Toxaphene (0.5 ppm) | • Chloroform (6 ppm) | • Nitrobenzene (2 ppm) |
| • Lead (5 ppm) | • 2,4-D (10 ppm) | • o-Cresol (200 ppm) | • Pyridine (5 ppm) |
| • Mercury (0.2 ppm) | • 2,4,5-TP (Silvex) (1 ppm) | • m-Cresol (200 ppm) | • Tetrachloroethylene (0.7 ppm) |
| • Selenium (1 ppm) | • Chlordane (0.03 ppm) | • p-Cresol (200 ppm) | • Trichloroethylene (0.5 ppm) |
| • Silver (5 ppm) | • Heptachlor (0.008 ppm) | • Cresols (200 ppm) | • 2,4,5-Trichlorophenol (400 ppm) |
| | • Pentachlorophenol (100 ppm) | • 1,4-Dichlorobenzene (7.5 ppm) | • 2,4,6-Trichlorophenol (2 ppm) |
| | • Hexachlorobenzene (0.13 ppm) | • 1,2-Dichloroethane (0.5 ppm) | • Vinyl chloride (0.2 ppm) |
| | | • 1,1-Dichloroethylene (0.7 ppm) | |
| | | • 2,4-Dinitrotoluene (0.13 ppm) | |

**Fig. (2).** Different class of hazardous materials present in waste.

**Table 2. Characteristics of hazardous wastes.**

| Hazardous | Characteristics |
| --- | --- |
| 1. Flammable/ Explosive | This type of waste may cause damage to the surroundings by Producing harmful gases at high temperature and pressure or by causing fire hazards. |
| 2. Oxidizng | Types of wastes that may yield oxygen and thereby cause or contribute to the combustion of other materials. |
| 3. Poisonous (Acute) | This waste has high potential to cause death, serious injury or to harm health if swallowed, inhaled or by skin contact. |
| 4. Infectious Substances | Hazardous wastes containing micro-organisms and their toxins, and responsible for diseases in animals or humans. |
| 5. Corrosives | These wastes are chemically active and may cause severe damage to the flora and fauna, or to the other materials by direct contact with them. |
| 6. Eco-toxic | These wastes may present immediate or delayed adverse impacts to the environment by means of bioaccumulation and/or toxic effects upon biotic systems. |
| 7. Toxic (Delayed or Chronic) | These wastes, if inhaled or ingested or if they penetrate the skin, may cause delayed or chronic effects, including carcinogenicity. |

*(Table 2) cont.....*

| **8.** Organic Peroxides | These are organic waste containing bivalent-O-O- structure and may undergo exothermic self-accelerating decomposition. |
|---|---|

**Ignitable Wastes**   **Corrosive Wastes**   **Reactive Wastes**   **Toxic Wastes**

Liquids having a flash point below 140F
a. alcohol solutions which contain more than 24% alcohol by volume
b. HPLC liquids (which often contain highly flammable solvents such as Acetonitrile)
c. glass-cleaning solvent rinses
Spontaneously combustible solids
a. metal powders
b. activated charcoal
Ignitable compressed gases
a. gas cylinders for lab burners (butane, propane, etc.)
Oxidizers
a. nitrate compounds
b. peroxide compounds
c. perchlorate compounds

Any liquid having a pH less than 2 or higher than 12.5
a. inorganic acids (hydrochloric, phosphoric, nitric, sulfuric, etc.)
b. organic acids (formic, acetic, lactic, etc.)
c. alkaline compounds (hydroxides, amines, etc.)

compounds which are normally unstable and readily undergo violent change without detonating
a. diethyl zinc
b. organometallic gases
compounds which will violently react with water
a. anhydrous metal salts
b. alkali metals
c. metal powders, shavings, or turnings compounds which form potentially explosive mixtures with water
a. calcium carbide
b. metal hydrides
c. chlorosilanes
compounds which are capable of detonation or explosive reaction
a. dry picric acid
b. Azo-, dinitro-, or trinitro compounds cyanide- and sulfide-bearing compounds

EPA Toxicity Characteristic List.

1  Heavy Metals
2  Pesticides Common
3  Organic Chemicals

**Characteristic Hazardous Waste**

**Fig. (3).** Characteristics of Hazardous materials present in wastes.

## QUANTITATIVE AND QUALITATIVE HAZARDOUS WASTE MANAGEMENT (HWM) IN INDIA

The quantity and quality of hazardous waste are influenced by different factors such as income levels [17], the sources [18, 19], population [20], social behaviors, religion, cast [22], industrial production [22], and the waste market [23, 24]. The amount of hazardous waste currently produced in India has grown from about 6 million to 48 million tons in 1947 and is expected to increase by 4.25 percent each year to 300 million tonnes by 2047 [25]. A survey of hazardous waste treatment NEERI in 59 cities (35, Metro Cities & 24 State Capitals: 2004/2005) by the Central Pollution Control Board (CPCB) was carried out. Table **3** reports the quantity and quality of dangerous waste data generated. In the past three decades, the per capita rate of hazardous waste has risen annually from 1 to 1.33% [26]. If this rate of growth continues, the generation of hazardous waste from less than

40,000 tons per year to more than 125,000 tons by 2030 will likely increase in India [27, 28].

**Table 3. Quantity and Quality of Municipal Hazardous Waste (MHW) generated at different cities in India 23.**

| S. No. | City | Population | Area (Km$^2$) | Waste Quantity (TPD) | Waste Generation Rate (kg/c/day) | Compostable (%) | Recyclables (%) | C/N Ratio | HCV* (Kcal/kg) | Moisture (%) |
|---|---|---|---|---|---|---|---|---|---|---|
| 1 | Kavaratti | 10119 | 4 | 3 | 0.30 | 46.01 | 27.20 | 18.04 | 2242 | 25 |
| 2 | Gangtok | 29354 | 15 | 13 | 0.44 | 46.52 | 16.48 | 25.61 | 1234 | 44 |
| 3 | Itanagar | 35022 | 22 | 12 | 0.34 | 52.02 | 20.57 | 17.68 | 3414 | 50 |
| 4 | Daman | 35770 | 7 | 15 | 0.42 | 29.60 | 22.02 | 22.34 | 2588 | 53 |
| 5 | Silvassa | 50463 | 17 | 16 | 0.32 | 71.67 | 13.97 | 35.24 | 1281 | 42 |
| 6 | Panjim | 59066 | 69 | 32 | 0.54 | 61.75 | 17.44 | 23.77 | 2211 | 47 |
| 7 | Kohima | 77030 | 30 | 13 | 0.17 | 57.48 | 22.67 | 30.87 | 2844 | 65 |
| 8 | Port Blair | 99984 | 18 | 76 | 0.76 | 48.25 | 27.66 | 35.88 | 1474 | 63 |
| 9 | Shillong | 132867 | 10 | 45 | 0.34 | 62.54 | 17.27 | 28.86 | 2736 | 63 |
| 10 | Shimla | 142555 | 20 | 39 | 0.27 | 43.02 | 36.64 | 23.76 | 2572 | 60 |
| 11 | Agartala | 18998 | 63 | 77 | 0.40 | 58.57 | 13.68 | 30.02 | 2427 | 60 |
| 12 | Gandhinagar | 195985 | 57 | 44 | 0.22 | 34.30 | 13.20 | 36.05 | 698 | 24 |
| 13 | Dhanbad | 199258 | 24 | 77 | 0.39 | 46.93 | 16.16 | 18.22 | 591 | 50 |
| 14 | Pondicherry | 220865 | 19 | 130 | 0.59 | 46.96 | 24.29 | 36.86 | 1846 | 54 |
| 15 | Imphal | 221492 | 34 | 43 | 0.19 | 60.00 | 18.51 | 22.34 | 3766 | 40 |
| 16 | Aizwal | 228280 | 117 | 57 | 0.25 | 54.24 | 20.97 | 27.45 | 3766 | 43 |
| 17 | Jammu | 369959 | 102 | 215 | 0.58 | 51.51 | 21.08 | 26.79 | 1782 | 40 |
| 18 | Dehradun | 424674 | 67 | 131 | 0.31 | 51.37 | 19.58 | 25.90 | 2445 | 60 |
| 19 | Asansol | 475439 | 127 | 207 | 0.44 | 50.33 | 14.21 | 14.08 | 1156 | 54 |
| 20 | Kochi | 595575 | 98 | 400 | 0.67 | 57.34 | 19.36 | 18.22 | 591 | 50 |
| 21 | Raipur | 605747 | 56 | 184 | 0.30 | 51.40 | 16.31 | 223.50 | 1273 | 29 |
| 22 | Bhubaneswar | 648032 | 135 | 234 | 0.36 | 49.81 | 12.69 | 20.57 | 742 | 59 |
| 23 | Tiruvanantapuram | 744983 | 142 | 171 | 0.23 | 72.96 | 14.36 | 35.19 | 2378 | 60 |
| 24 | Chandigarh | 808515 | 114 | 326 | 0.40 | 57.18 | 10.91 | 20.52 | 1408 | 64 |
| 25 | Guwahati | 809895 | 218 | 166 | 0.20 | 53.69 | 23.28 | 17.71 | 1519 | 61 |
| 26 | Ranchi | 847093 | 224 | 208 | 0.25 | 51.49 | 9.86 | 20.23 | 1060 | 49 |
| 27 | Vijaywada | 851282 | 58 | 374 | 0.44 | 59.43 | 17.40 | 33.90 | 1910 | 46 |
| 28 | Srinagar | 898440 | 341 | 428 | 0.48 | 61.77 | 17.76 | 22.46 | 1264 | 61 |
| 29 | Madurai | 928868 | 52 | 275 | 0.30 | 55.32 | 17.25 | 32.69 | 1813 | 46 |
| 30 | Coimbatore | 930882 | 107 | 530 | 0.57 | 50.06 | 15.52 | 45.83 | 2381 | 54 |
| 31 | Jabalpur | 932484 | 134 | 216 | 0.23 | 58.07 | 16.61 | 28.22 | 2051 | 35 |
| 32 | Amritsar | 966862 | 77 | 438 | 0.45 | 65.02 | 13.94 | 30.69 | 1836 | 61 |
| 33 | Rajkot | 967476 | 105 | 207 | 0.21 | 41.50 | 11.20 | 52.56 | 687 | 17 |

*(Table 3) cont.....*

| 34 | Allahabad | 975393 | 71 | 509 | 0.52 | 35.49 | 19.22 | 19.00 | 1180 | 18 |
|----|-----------|--------|-----|------|------|-------|-------|-------|------|----|
| 35 | Vishakhapatnam | 982904 | 110 | 584 | 0.59 | 45.96 | 24.20 | 41.70 | 1602 | 53 |
| 36 | Faridabad | 1055938 | 216 | 448 | 0.42 | 42.06 | 23.31 | 18.58 | 1319 | 34 |
| 37 | Meerut | 1068772 | 142 | 490 | 0.46 | 54.54 | 10.96 | 19.24 | 1089 | 32 |
| 38 | Nashik | 1077236 | 269 | 200 | 0.19 | 39.52 | 25.11 | 37.20 | 2762 | 62 |
| 39 | Varanasi | 1091918 | 80 | 425 | 0.39 | 45.18 | 17.23 | 19.40 | 804 | 44 |
| 40 | Jamshedpur | 1104713 | 64 | 338 | 0.31 | 43.36 | 15.69 | 19.69 | 1009 | 48 |
| 41 | Agra | 1275135 | 140 | 654 | 0.51 | 46.38 | 15.79 | 21.56 | 520 | 28 |
| 42 | Vadodara | 1306227 | 240 | 357 | 0.27 | 47.43 | 14.50 | 40.34 | 1781 | 25 |
| 43 | Patna | 1366444 | 107 | 511 | 0.37 | 51.96 | 12.57 | 18.62 | 819 | 36 |
| 44 | Ludhiyana | 1398467 | 159 | 735 | 0.53 | 49.80 | 19.32 | 52.17 | 2559 | 65 |
| 45 | Mumbai | 1437354 | 286 | 574 | 0.40 | 52.44 | 22.33 | 21.58 | 1421 | 43 |
| 46 | Indore | 1474968 | 130 | 557 | 0.38 | 48.97 | 12.57 | 29.30 | 1437 | 31 |
| 47 | Nagpur | 2052066 | 218 | 504 | 0.25 | 47.41 | 15.53 | 26.37 | 2632 | 41 |
| 48 | Lucknow | 2185927 | 310 | 475 | 0.22 | 47.41 | 15.53 | 21.41 | 1557 | 60 |
| 49 | Jaipur | 2322575 | 518 | 904 | 0.39 | 45.50 | 12.10 | 43.29 | 834 | 21 |
| 50 | Surat | 2433835 | 112 | 1000 | 0.41 | 56.87 | 11.21 | 42.16 | 990 | 51 |
| 51 | Pune | 2538473 | 244 | 1175 | 0.46 | 62.44 | 16.66 | 35.54 | 2531 | 63 |
| 52 | Kanpur | 2551337 | 267 | 1100 | 0.43 | 47.52 | 11.93 | 27.64 | 1571 | 46 |
| 53 | Ahmedabad | 3520085 | 191 | 1302 | 0.37 | 40.81 | 11.65 | 29.64 | 1180 | 32 |
| 54 | Hyderabad | 3843585 | 169 | 2187 | 0.57 | 54.20 | 21.60 | 25.90 | 1969 | 46 |
| 55 | Bangalore | 4301326 | 226 | 1669 | 0.39 | 51.84 | 22.43 | 35.12 | 2386 | 55 |
| 56 | Chennai | 4343645 | 174 | 3036 | 0.62 | 41.34 | 16.34 | 29.25 | 2594 | 47 |
| 57 | Kolkata | 4572876 | 187 | 2653 | 0.58 | 50.56 | 11.48 | 31.81 | 1201 | 46 |
| 58 | Delhi | 10306452 | 1483 | 5922 | 0.57 | 54.42 | 15.52 | 34.87 | 1802 | 49 |
| 59 | Greater Mumbai | 11978450 | 437 | 5320 | 0.45 | 62.44 | 16.66 | 39.04 | 1786 | 54 |

Moreover, some per capita cities include Chennai, Port Blair, Kochi, Pondicherry, Vishakhapatnam, Jammu, Kolkata, Hyderabad and Delhi, due to higher standard of living, rapid economic growth and higher urbanization levels. These cities have a high per capita generation rate. However, in other cities like Rajkot, Nasik, Guwahati, and Imphal, the per capita rate of generation is observed to be small. Differing distribution and amounts of hazardous waste from different cities is attributed to the demographic size, different consumption patterns, and the indices for the production of waste, the socio-economic and cultural scales and, to a large extent, the effects of consumption patterns [29, 30]. The differences found in different cities are compostable (40%-60%) and inert (30%-50%) municipal hazardous solid waste (MHSW). The main part of MHSW consisted of organic materials (44%), compared with recyclables such as metal [31], plastics, glass and paper. MHSW contributed most to recyclable materials [21]. Finally, food waste is a major component of all the MHSW constituents with a high percentage [32, 33].

## COMPOSITION OF MUNICIPAL HAZARDOUS WASTE

MHW's composition depends on a variety of factors including cultural traditions, dietary practices [34 - 36] and climate and incomes [37, 38]. The cities producing more MHW per day per capita have been found in India and other developing countries and their waste has higher sections of packaging and recyclable waste, while the recyclable wastes are very small for developing countries and the proportion is compostable (Fig. **5**). The partially degradable MHWs are discarded napkins, wood or sludge, sanitary remains or non-degradable products, such as plastics, leather, rubbers, bottles, metals and oil ashes, such as carbon, briquettes and wood, electronic waste or dust [33]. The highest percentage (40 to 70 percent) in India of MHW is composed of organic materials that are capable of keeping high moisture levels in food waste, which contributes to the MHW [35 - 38]. The ratio of C/N ranges from 800 to 1000 kcal/kg [39, 40].

## BIOMEDICAL WASTE MANAGEMENT (BWM) IN INDIA

The BMW Rules, 2016 provide the procedure more specificity. The Rules apply to hospitals, nursing homes, clinics, dispensaries, animal houses, veterinary institutions, blood banks, clinical establishments, pathological laboratories, research or educational institutions, medical or surgical camps, health camps, blood donation camps, vaccination camps, forensic laboratories, research labs and first aid rooms of schools. It has been specified that the BMW Rules, 2016 do not apply to hazardous wastes, radioactive wastes, hazardous chemicals, municipal solid wastes, hazardous microorganisms, genetically engineered microorganisms and cells, e-wastes, lead acid batteries, which are regulated by the other applicable Rules [41 - 43].

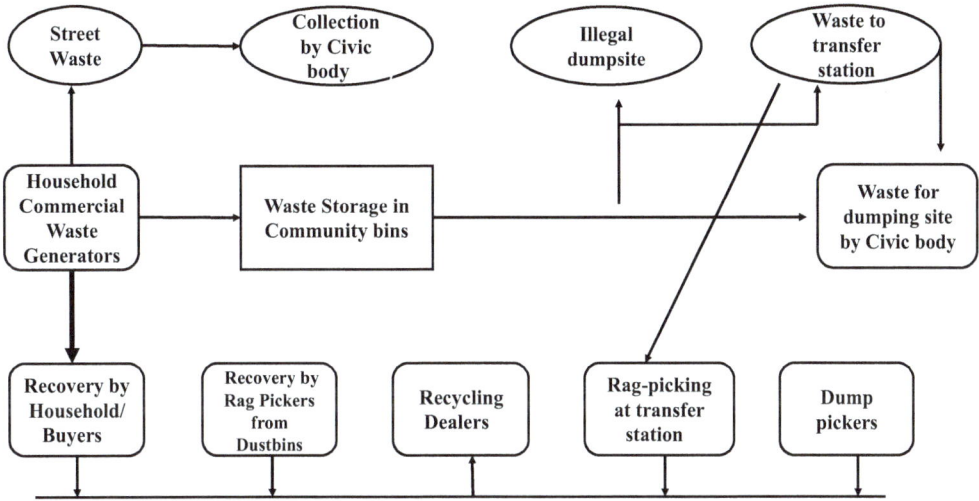

**Fig. (4).** Schematic Flow Chart of Common MHW Management Process (**Source: Joseph, 2002**).

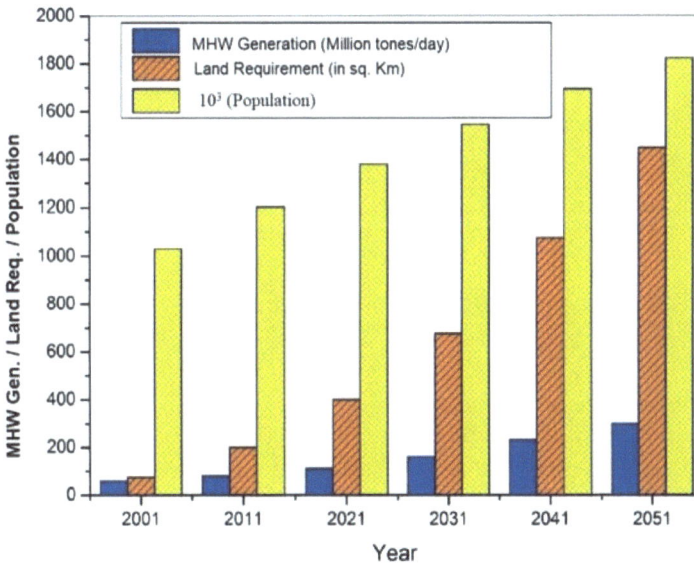

**Fig. (5).** Prediction Plot for MHW generation, land requirement, and population from 2001 to 2051 (**Source:** Joshi & Ahmed 2016).

## Biomedical Waste Management Rules, 2016

- *Occupier:* a person having administrative control over the institution and the premises generating biomedical waste including hospital, nursing home, clinic, dispensary, veterinary institution, animal house, pathological laboratory, blood bank, health care facility and clinical establishment [43].

- *Operator:* a person who owns or controls a common biomedical waste treatment facility for the collection, reception, storage, transport, treatment, disposal or any other form of handling of biomedical waste [43].

- *Prescribed authority:* The State Pollution Control Board in respect of a State and Pollution Control Committee in respect of a Union territory [43].

- *Treatment and disposal facility:* facility wherein treatment, disposal of biomedical waste or processes incidental to such treatment and disposal is carried out, and includes common bio-medical waste treatment facilities [43].

It is compulsory that each State Government or Union Territory Administration shall, under the chairmanship of the respective Health Secretary, set up an Advisory Committee for the respective State or Union territory to oversee the implementation of the rules in the respective State. Therefore, under the chairmanship of District Collector or District Magistrate or Deputy Commissioner or Additional District Magistrate, any State Government or Union Territory Administration shall set up a district-level monitoring committee to ensure compliance with the provisions of those rules in health centers and in specific biomedical waste treatment and disposal facilities. The Monitoring Committee shall submit its report to the State Advisory Committee once every 6 months, with a copy approved by the State Pollution Control Board or Pollution Management Committee concerned for further action required [41 - 43]. Table **4** indicates the significant differences between BMWM regulations 2016 and 1998. A business must buyback policies and handle them according to the approved global guidelines for biological waste not reported in the current regulations of the country.

### Risks Associated with Biomedical Hazardous Waste

- The risk to healthcare workers and waste handlers: Contaminated sharps that are improperly processed carry the greatest risk of contamination associated with hospital waste. Pathogens that may be aerosolized during the compacting, grinding or shredding process associated with some medical waste collection or disposal procedures often face potential safety risk to medical waste handling. Physical (injury) and environmental risks are also related to the elevated running temperatures of incinerators and vapor sterilizers and harmful gasses that are discharged into the environment after waste disposal [44].

- The risk to the public: Public impacts are confined to esthetic degradation of the environment from careless disposal and the environmental impact of improperly operated incinerators or other medical waste treatment equipment. In patients, there could be an increased risk of nosocomial infections due to inadequate handling of the waste. Improper management of waste can lead to altered microbial ecology and the spread of antibiotic resistance [44].

**Table 4. The salient differences between biomedical waste management rules 2016 and biomedical waste rules 1998 41.**

| Points | Biomedical Waste Management Rules, (2016) | Biomedical Waste, 1998 |
|---|---|---|
| **Duties of Occupier** | Duties of occupier are delineated better.<br>There is pretreatment by: disinfection and sterilisation on-site of infectious lab waste blood bags as per WHO guidelines.<br>Occupier ensures non-chlorinated plastic bags, gloves, blood bags within 2y of notification.<br>Occupier ensures liquid waste is segregated at source by pretreatment and Effluent treatment plant (ETP) is mandatory.<br>Occupier ensures to maintain BMWM register daily and on website monthly.<br>Annual report should be made available on web-site within 2 years.<br>The occupier ($\geq 30$ bedded) establishes BMWM committee.<br>Records of equipments, training, health checkup, immunization compulsory. | Duties of occupier not delineated better.<br>No pretreatment of waste on site.<br>Chlorinated plastic bags, gloves, blood bags were recommended.<br>ETP not mandatory.<br>The details of records not mandatory.<br>The annual report need not to be posted on website.<br>BMWM committee not compulsory.<br>Records not compulsory to maintain. |
| **Duties of CBMWTF** | Duties are delineated better.<br>Occupier has to establish Bar coding and GPS by first year and ensure occupational safety of all its HCWs by TT and HBV vaccination.<br>Reporting of accidents and maintenance of records of equipments, training, health checkup. | Duties are not delineated better.<br>Barcoding and GPS not documented and vaccinations for HCWs not documented.<br>Records not documented. |
| **Accident Reporting** | Major accidents are reported to authorities and in annual report. | No specific reporting of accidents. |
| **Schedule I** | Yellow | Yellow |
| **Deep burial** | Deep burial is an option for remote and rural areas. | Deep burial allowed in villages and towns with less than 5 lakhs population. |

*(Table 4) cont.....*

| Chemical treatment | Chemicals treatment using 10% hypochlorite solution. | Chemical treatment:1% hypochlorite. |
|---|---|---|
| **Fetus** | Fetus less than the age of viability is to be treated as human anatomical waste. | No demarcation of fetus mentioned. |
| **Drugs** | Antibiotics and other drugs and solid chemical waste suggested for incineration. Cytotoxic drugs: incineration up to 1200 C and return back to supplier. | All drugs to be discarded in black bag. For cytotoxic drugs destruction and drugs disposal in secured landfills. |
| **Liquid infected waste** | Effluent treatment plant is mandatory and effluent to conform to standards mentioned. | For liquid waste chemical treatment and discharge in to drains to conform to effluent standards mentioned. |
| **Microbiology and biotechnology waste Infected plastics, sharps, glass** | Pretreatment of infectious waste is as per WHO guidelines: log6 and log 4 reduction. The infected plastics and sharps after mutilation go in for red bag and white container respectively and are sent to authorized recyclers. The glass articles are discarded in cardboard box with blue marking. | Pretreatment not mandatory. Infected plastics, metal sharps, glass go in blue container with disinfectant and local autoclaving/microwaving/incineration is recommended. |
| **Recycling** | A focus on recycling of plastic, sharps, glass to authorized recyclers. | Recycling of plastics, glass to authorized recyclers not mentioned. |

## Basic Steps of Hazardous Biomedical Waste Management

Medical waste should be treated according to the type and characteristics of the waste. The waste should be treated at every stage, from collection to disposal, so that waste management is more effectively successful. The components of an effective waste management system are as follows: waste survey, segregation, accumulation and storage, transportation, treatment [44].

## Waste Survey

The Waste survey should categorize and quantify the waste generated. It should determine the points of generation, the type of waste at each stage, and the level of generation and sanitization within the hospital. This helps to regulate the proper method of disposal [44].

# WASTE SEGREGATION

In this process, placing different kinds of wastes in different containers or coded bags at the point of generation. It helps to reduce the higher amount of infectious waste as well as treatment costs. Segregation also helps to reduce the spread of infection to other health care workers [44].

## Waste Accumulation and Storage

It occurs between the various stage of waste generation and different sites of waste treatment and disposal. While accumulation refers to the temporary holding of small quantities of waste near the specific stage of generation, storage of waste is characterized by longer holding periods and large waste quantity. Storage areas are generally located near where the waste is treated [44].

## Waste Transportation

Medical waste is not treated on site. Untreated medical waste must be transported from the generation facility to another site for treatment and disposal [44].

# WASTE TREATMENT

In this process, the treatment of mainly required to disinfect for decontaminating the waste. After such treatment, the residue can be handled safely, transported, and stored [44].

# SAFELY DISPOSAL AND PREVENTIVE METHODS OF HAZARDOUS WASTE MANAGEMENT IN INDIA

The different methods of treatment required for the management of hazardous waste are:

- Landfilling.
- Incineration.
- Recycling and reuse.

## Landfilling

It is one of the most frequently used methods of waste disposal. Trenches are constructed on flat surfaces in the settlements. The earth is excavated from the pits, and a large amount of dirt is covered in the waste material. Certain facilities such as impermeable plastic or clay liner, a collection basin that collects and translates leachate to wastewater plants, are equipped with modern techniques like safe disposal. The processes of degradation in sites are very complicated and last for a long time [45, 46].

## Hazards of landfilling

- Leaking landfills
- Leachate contaminating soil and groundwater
- Chemical reactions
- Vaporization
- Uncontrolled fires

Therefore, landfilling does not tend to be an environmentally sound treatment method of uncertain comportment on a waste site.

## Incineration

The waste material is burned in high-temperature specially designed combustions (900-1000 °C) through a controlled and complete combustion process. The benefit of hazardous waste incineration is a decrease in waste size and the use of fuel energy content. Incineration disadvantages include emission into the environment of flue gas fuelling substances and the high quantity of gas waste cleaning and combustion [45]. The global emissions of cadmium and mercury were significantly contributed from hazardous waste incineration facilities. For contrast, heavy metals will not be released into the air and converted into waste kill or exhaust gas which can be disposed of again in the forest. Consequently, incineration of hazardous waste will raise these emissions if steps such as the elimination of heavy metals are not taken [45].

## Incineration Hazards

- Dioxin formation.
- Heavy metal contamination.
- Contaminated slag, fly ash and flue gases.
- Health and safety hazards.

## Recycling of Hazardous Waste

End-of-life electronics plastics management options are three types of recycling.

- Chemical recycling.
- Mechanical recycling.
- Thermal recycling.

Any recycling process includes the removal of different parts of e-waste that contain hazardous substances, such as PCB, Hg, plastic sorting, CRT, ferrous or nonferrous metal segregation and printed circuit boards. The extraction of precious metals like copper, lead, and gold obtain on strong acids. If suitable technology is used, the value of recycling from the element could be much higher. Without mask and technological expertise, recycles work in poorly ventilated enclosures and result in exposure to dangerous and slow toxic substances. It is possible to reuse displays and CRTs, keyboards, compact disks and mobiles, desktop computers, scanners, Processors, memory chips, wires and cables [45 - 47].

## Hazardous Effects Due to Recycling

- Potential threat to human health and the environment.
- Lead causes damage to the central and peripheral nerve system, blood system and kidneys in humans.
- Mercury impacts brain functioning and development.

Recycling is therefore the best option for hazardous waste management [45].

## REUSE

Following minor modifications to the original running equipment, it is either used or modified. It is often used for electronic devices like laptops, mobile devices, *etc*. After re-filling, inkjet cartridges are also used. This method also reduces risky waste generation volumes [45].

## EMERGING TECHNOLOGIES

Plasma pyrolysis, alkaline hydrolysis, over-heated steam, ozone, and hope are new technologies [48, 53]. Plasma pyrolysis uses ionized gas in the plasma state for the conversion from plasma or electrodes with minimal or no air of electrostatic energy to temperatures of several thousand degrees. It is used to decompose toxic infection, bacterial, plastic, industrial and pharmaceutical waste. It is clean, eco-friendly, energy-saving and low-carbon dioxin and furan emissions. Clean alloyed layer production, which can be employed in building materials and products with value-added like metals. Their disadvantages include large initial investment costs, carbon dioxide pollution, large electrical inputs and the frequency of maintenance of highly corrosive plasma fire [45, 48, 49]. In medicines, water and air therapy, in general, ozone ($O_3$) can be used. It is a strong oxidizer and is more compact in form ($O_2$). To expose waste to the bactericidal agent, Ozone systems require shredders and mixers. Daily research should be carried out to ensure the quality of microbial inactivation is met [45 - 48].

## Promession

Freezing with liquid nitrogen and mechanical vibration is included to disintegrate cadavers before the burial. It includes freezing. This process accelerates the decomposition, reduces weight and volume and enables metal parts to be recovered [45 - 49].

## Alkaline Hydrolysis

This turns parts of the body, samples and bodies into a decontaminated aqueous solution that removes prion infected materials, dangerous chemicals and waste. Once the waste is packed into the container and into the enclosed tank, alkali and water are applied and combined at a temperature of 127 °C or lower. After 6–8 hours of digestion, by-products are made up of bone and tooth mineral elements, amino acid solution, sugar, soap and salts. It may also kill chemotherapy and cytotoxic agents, as well as other aldehydes widely used in hospitals, such as formaldehyde and glutaraldehyde [45, 50, 51].

## Nanotechnology

It is used to clean the air and improve air quality indoors and includes a large-spectrum photo buffer and is bactericidal and fungicidal. This uses light-energy to produce hydroxyl and superoxide anion ($O_2^-$) which decompose and oxidize carbon dioxide and water toxic pollutants [45 - 52].

## Photocatalysis

It is a new technology that uses solar or ultraviolet rays to disinfect microbes and decontaminate waste-water antibiotics at their point of origin, for the disinfestation and decontamination of hospitable wastewater. The software is effective and cost-efficient [45 - 53].

## Membrane Bioreactors

This combines bioactivated sludge with a membrane filtering step for the isolation of sludge water. Different membrane bioreactor forms, such as aerobic MBR, anaerobic MBR, organic pollutant MBR [45 - 55] are accessible. Many new technologies for BMW destruction developments include gas-phase chemistry reduction, base-catalyzed decomposition, supercritical liquid corrosion, sodium reduction, screening, overheated vaporeforming, Fe-TAML/peroxide (pharmaceutical waste) disposal, biodegradation (mealworms and food plastic bacteria), mechochemical processing, acoustic engineering, electrical technology and so on. These new technologies are not ready for routine use of health care waste [45, 56].

## CHALLENGES FOR WASTE MANAGEMENT IN INDIA

It is imperative that people become more environmentally aware and engage residents in isolating the waste from source, collecting door to door, and disposing of waste treatment sites. This is a critical stage in the entire MSWM process, contributing to the handling of solid waste that results in ultimate success [57, 58]. But in India, the current scenario shows that there is almost no waste segregation at source. And in addition, the residents throw trash improperly due to the lack of coordination between their citizens and the lack of planned cities in India. In the end, community containers are not available in the nearby neighborhood and the number of ULB employees is insufficient according to the population living there [58]. Furthermore, groups of officers and qualified personnel need to be created for ULBs with MSWM expertise. Proper training and practical experiments would enable it to identify and take appropriate action at the implementation level bottlenecks [59]. There is less dialog between the government of the Central and State. Delay in providing State to Central information delays the corresponding level on the ground. The key impediment to this lack of coordination on the specific action plan and the inadequate ULB policy at executing level [57]. Specifically, an appropriate technological solution through Public-Private Partnerships (PPP) is necessary, so that the performance of the PPP would entail the development of good public governance in compliance with an established regulatory framework and adequate financial support. The growth and supply of skilled labor are effective role-playing with current and good practices for SWM, financial incentives to identify new technically viable approaches, effective and rapid ULBs decision to incorporate them seamlessly are key challenges [57, 58]. India continues to struggle to make a successful waste-to-energy project. It is necessary to import technologies that are cost-effective and proven. In addition, properly identified and segregated waste must be delivered, according to its requirements, to waste-to-energy plants [60, 61]. To improve productivity to MSW recycling and the isolation of sources, recyclists can be active in the coordinated market. This immense potential has, however, been ignored [61, 62] because of the absence of recycling industries and the acceptance of society.

## CONCLUSIONS

In summary, it can be said that in developing countries, production costs are often given priority over the best available technologies and this results in more waste generation. It is difficult to develop alternative technologies for the total elimination of hazardous waste generation, but we can take action to use renewable options such as solar energy, wind energy in the production process, instead of using power for processing, as this can generate fewer emissions of hazardous waste by going towards green — non-pollution. With this, we can formulate policies and strategies to prioritize the reduction and minimization of waste rather than mere disposal. The remediation plan needs to focus on the 'polluter costs' concept, with polluters being asked to pay the fine as well as the cost of cleaning up the waste. Industries that produce waste should be blacklisted regularly. Where polluters are not traceable, the State Waste Control Board/Pollution Control Committee (SPCB/PCC) need to set up a separate fund for remediation. With this in opinion, Waste Exchange Banks/Collection Centers should be developed to provide information on waste as well as on types of waste and waste management methods, to provide information on waste and to promote re-use, recovery and recycling technologies that enhance the quality of the recovery of resources.

Now, India is growing economic growth and industrialization present significant risks to the environment and public health due to the subsequent generation and inadequate management of hazardous waste. However, a national campaign launched in India by Prime Minister Narendra Modi for 2014-2020 aimed at clearing up roads, streets, public toilets and other private and public infrastructure in Indian cities and rural areas is currently focusing strongly on the Swachh Bharat Mission (SBM) or Swachh Bharat Abhiyan (SBA). In this mission, the public is increasingly aware and legislation is implemented to enhance effective hazardous waste management at different levels. The ambitious initiatives by Indian Prime Minister Narendra Modi are a chance to demonstrate to the world that India is willing to address future issues such as climate change and to identify reliable global standards for the control, ecological and occupational health of hazardous waste. Swachh Bharat Mission has also played a vital and important role in raising public awareness of the waste-related health issue.

## LIST OF ABBREVIATIONS

| | |
|---|---|
| **HWM** | Hazardous waste management |
| **ULB** | Local urban bodies |
| **MHWM** | Municipal Hazardous waste management |
| **MOEF** | Ministry of Environment and Forests |
| **CRB** | Crop residue burning |
| **HW** | Hazardous waste |
| **CPCB** | Central Pollution Control Board |
| **SPCB** | State Pollution Control Boards |
| **CPCB** | Pollution control board |
| **MHSW** | Municipal hazardous solid waste |
| **MHW** | Municipal hazardous waste |
| **HCF** | Health care facility |

| | |
|---|---|
| **CBMW** | Common biomedical waste management |
| **CBMW** | Common biomedical waste management |
| **HCW** | Healthcare workers |
| **ETP** | Effluent treatment plant |

## CONSENT FOR PUBLICATION

Not applicable.

## CONFLICT OF INTEREST

The authors confirm that the contents of this chapter have no conflict of interest.

## ACKNOWLEDGEMENTS

Declared none.

## REFERENCES

[1]     Khan DJ, Kaseva ME, Mbuligue SE. Hazardous waste issues in developing countries

[2]     Sureka J, Saranya Devi K. Synthesis of silver nano particle from gymnema sylvestre: Study of antimicrobial and antioxidant activity. Int J Creat Res Thoughts 2018; 6(1): 897-902.

[3]     PPCB (Punjab Pollution Control Board). 2010. http://www.ppcb.gov.in/Attachments/Annual%20 Reports/AR201011.pdf

[4]     Narayana T. Municipal solid waste management in India: From waste disposal to recovery of resources? Waste Manag 2009; 29(3): 1163-6.
[http://dx.doi.org/10.1016/j.wasman.2008.06.038] [PMID: 18829290]

[5]     Biswas AK, Kumar S, Babu SS, Bhattacharyya JK, Chakrabarti T. Studies on environmental quality in and around municipal solid waste dumpsite. Resour Conserv Recycling 2010; 55: 129-34.
[http://dx.doi.org/10.1016/j.resconrec.2010.08.003]

[6]     MOE&F. E-waste (Management) Rules. Delhi: Ministry of Environment & Forests 2016. Retrieved from  http://www.moef.nic.in

[7]     MoEF & CC (The Ministry of Environment, Forests and Climate Change). The Hazardous Waste (Management and Handling) Rules; MoEF & CC: New Delhi, India 2008.

[8]     Kumar S, Smith SR, Fowler G, *et al.* Challenges and opportunities associated with waste management in India. R Soc Open Sci 2017; 4(3): 160764.
[http://dx.doi.org/10.1098/rsos.160764] [PMID: 28405362]

[9]     ELI (Environmental Law Institute). Enforcing Hazardous Waste Rules in India: Strategies and Techniques for Achieving Increased Compliance. Washington, DC, USA: ELI 2014.

[10]    Indo-UK Seminar Report 2015. http://www.neeri.res.in/Short%20Report_Indo-UK%20Seminar%20 (25-27th%20March%202015.pdf)

[11]    Kumar S, Smith SR, Fowler G, *et al.* Challenges and opportunities associated with waste management in India. R Soc Open Sci 2017; 4: 160764.

[12]    The Associated Chambers of Commerce of India (ASSOCHAM). PricewaterhouseCoopers (PwC) Waste Management in India-Shifting Gears. New Delhi, India: ASSOCHAM 2017.

[13]    Ahuja AS, Abda SA. Industrial hazardous waste management by government of Gujarat. Res Hub Int Multidiscip Res J 2015; 2: 1-11.

[14]    Ozaki H, Sharma K, Phanuwan C, Fukushi K, Polprasert C. Management of hazardous waste inThailand: Present situation and future prospects. J Mater Cycles Waste Manag 2004; 5: 31-8. [http://dx.doi.org/10.1007/s101630300006]

[15]    EBTC (European Business and Technology Centre). Waste Management in India: A Snap Shot http://ebtc.eu/index.php/knowledge-centre/publications/environment-publications/

[16]    Tudor, T., Vaccari, M. Preface: Special issue on innovative processes and technologies for the management of hazardous waste. Environments 2018; 5(10): 106. [http://dx.doi.org/10.3390/environments5100106]

[17]    Minghua Z, Xiumin F, Rovetta A, *et al.* Municipal solid waste management in Pudong New Area, China. Waste Manag 2009; 29(3): 1227-33. [http://dx.doi.org/10.1016/j.wasman.2008.07.016]

[18]    Jha AK, Sharma C, Singh N, Ramesh R, Purvaja R, Gupta PK. Greenhouse gas emissions from municipal solid waste management in Indian mega-cities: A case study of Chennai landfill sites. Chemosphere 2008; 71(4): 750-8. [http://dx.doi.org/10.1016/j.chemosphere.2007.10.024] [PMID: 18068211]

[19]    Hoornweg D, Bhada-Tata P. What a waste: a global review of solid waste management. Washington DC, USA: World Bank 2012.

[20]    Thanh NP, Matsui Y, Fujiwara T. Assessment of plastic waste generation and its potential recycling of household solid waste in Can Tho City, Vietnam. Environ Monit Assess 2011; 175(1-4): 23-35. [http://dx.doi.org/10.1007/s10661-010-1490-8] [PMID: 20490914]

[21]    Gupta S, Mohan K, Prasad RK, Kansal A. Solid waste management in India: Options and opportunities. Resour Conserv Recycling 1998; 24(2): 137-54. [http://dx.doi.org/10.1016/S0921-3449(98)00033-0]

[22]    Basiago AD. Economic, social, and environmental sustainability in development theory and urban planning practice, 19 ed., Kluwer Academic Publishers, Boston. Manufactured in the Netherlands. Environmentalist 1999.

[23]    Central Pollution Control Board (CPCB), management of municipal solid waste. New Delhi, India: Ministry of Environment and Forests 2004.

[24]    Central Public Health and Environmental Engineering Organization (CPHEEO) 2000.

[25]    Sharholy M, Ahmad K, Mahmood G, Trivedi RC. Development of prediction models for municipal solid waste generation for Delhi city 2006.

[26]    Shekdar AV. Municipal solid waste management–the Indian perspective. J Ind Assoc Environ Manag 1999; 26(2): 100-8.

[27]    Srishti, 2000 Fifth Srishti survey of medical waste disposal practices in health care units of Delhi New Delhi

[28]    Minghua Z, Xiumin F, Rovetta A, *et al.* Municipal solid waste management in Pudong New Area, China. Waste Manag 2009; 29(3): 1227-33. [http://dx.doi.org/10.1016/j.wasman.2008.07.016] [PMID: 18951780]

[29]    Bernache-Pérez G, Sánchez-Colón S, Garmendia AM, Dávila-Villarreal A, Sánchez-Salazar ME. Solid waste characterisation study in the Guadalajara Metropolitan Zone, Mexico. Waste Manag Res 2001; 19(5): 413-24. [http://dx.doi.org/10.1177/0734242X0101900506] [PMID: 11954727]

[30]    Buenrostro O, Bocco G, Vence J. Forecasting generation of urban solid waste in developing countries—a case study in Mexico, *Journal of the Air & Waste Management Association, 51(1),*2001,

86-93. R. Mohee, S. Mauthoor, Z. M. Bundhoo, G. Somaroo, N. Soobhany, and S. Gunasee, Current status of solid waste management in small island developing states: a review. Waste Manag 2015; 43(1): 539-49.

[31]   Ojeda-Benítez S, Beraud-Lozano JL. The municipal solid waste cycle in Mexico: final disposal. Resour Conserv Recycling 2003; 39(3): 239-50.
[http://dx.doi.org/10.1016/S0921-3449(03)00030-2]

[32]   Agamuthu P, Fauziah SH, Khidzir KM, Aiza AN. Sustainable waste management-Asian perspectives Proceedings of the international conference on sustainable solid waste management. 15-26.

[33]   Bhat RA, Kamili AN, Bandh SA. Characterisation and composition of municipal solid waste (MSW) generated in Yusmarg: a health resort of Kashmir valley. A Glance at the World. Waste Manag 2013; 33(1): 774-7.

[34]   Satterthwaite D, McGranahan G, Tacoli C. Urbanization and its implications for food and farming. Philos Trans R Soc Lond B Biol Sci 2010; 365(1554): 2809-20.
[http://dx.doi.org/10.1098/rstb.2010.0136] [PMID: 20713386]

[35]   Agamuthu P, Fauziah SH, Khidzir KM, Aiza AN. Sustainable waste management-Asianperspectives Proceedings of the international conference on sustainable solid waste management. 15-26.

[36]   Bhat RA, Kamili AN, Bandh SA. Characterisation and composition of municipal solid waste (MSW)generated in Yusmarg: a health resort of Kashmir valley. A Glance at the World. Waste Manag 2013; 33(1): 774-7.

[37]   Gupta S, Mohan K, Prasad RK. Gupta, and A. Kansal, Solid waste management in India: options andopportunities. Resour Conserv Recycling 1998; 24(2): 137-54.
[http://dx.doi.org/10.1016/S0921-3449(98)00033-0]

[38]   Kumar S, Bhattacharyya J K, Vaidya AN, Chakrabarti T, Devotta S, Akolkar A B. Assessment ofthe status of municipal solid waste management in metro cities, state capitals, class I cities, and class IItowns in India: An insight. Journal of waste management, 2009; 29(2): 883-95.

[39]   Sharholy M, Ahmad K, Mahmood G, Trivedi RC. Municipal solid waste management in Indian cities - A review. Waste Manag 2008; 28(2): 459-67.
[http://dx.doi.org/10.1016/j.wasman.2007.02.008] [PMID: 17433664]

[40]   Kaushal RK, Varghese GK, Chabukdhara M. Municipal solid waste management in India-current state and future challenges: a review. Int J Eng Sci Technol 2012; 4(4): 1473-89.

[41]   Capoor MR, Bhowmik KT. Current perspectives on biomedical waste management: Rules, conventions and treatment technologies. Indian J Med Microbiol 2017; 35(2): 157-64.
[PMID: 28681801]

[42]   Bio-Medical Waste Management Rules. 2016.

[43]   Kharat Dr. Biomedical Waste Management Rules, 2016: A review. International of advanced. Res Dev 2016; 1: 48-51.

[44]   Hegde V, Kulkarni R, Ajantha G. Biomedical waste management. J Oral Maxillofac Pathol 2007; 11(1): 5.
[http://dx.doi.org/10.4103/0973-029X.33955]

[45]   Annamalai J. Occupational health hazards related to informal recycling of E-waste in India: An overview. Indian J Occup Environ Med 2015; 19(1): 61-5.
[http://dx.doi.org/10.4103/0019-5278.157013] [PMID: 26023273]

[46]   Ramesh S, Joseph K. Electronic waste generation and management in an Indian city. J Indian Assoc Environ Manage 2006; 33: 100-5.

[47]   Radha G. A Study of the Performance of the Indian IT Sector 2002. http://www.oldsite.nautilus.org/ archives/cap/reports/ IndiaExecSummary.pdf

[48]    Emmanuel J. Non-incineration medical waste treatment technologies. 2001.

[49]    Nema SK, Ganeshprasad KS. Plasma pyrolysis for medical waste. Curr Sci 2002; 83: 271-8.

[50]    European Commission Scientific Steering Committee. Updated Opinion and Report on a Treatment of Animal Waste by Means of High Temperature and High Pressure Alkaline Hydrolysis 2003.http://www.ec.europa.eu/food/fs/sc/ssc/out297_en.pdf

[51]    Qu X, Alvarez PJ, Li Q. Applications of nanotechnology in water and wastewater treatment. Water Res 2013; 47(12): 3931-46.
       [http://dx.doi.org/10.1016/j.watres.2012.09.058] [PMID: 23571110]

[52]    Alrhmoun M. https://www.tel.archives-ouvertes.fr

[53]    Judd S, Judd C, Eds. Principles and Applications of Membrane Bioreactors for Water and Wastewater Treatment System. 2nd ed., United Kingdom: Elsevier 2011.

[54]    EPA. Reference Guide to Non-Combustion Technologies for Remediation of Persistent Organic Pollutants in Stockpiles and Soil 2005.

[55]    Joshi & Ahmed. Status and challenges of municipal solid waste management in India: A review. Cogent Environ Sci 2016; 2: 1139434.

[56]    ELI (Environmental Law Institute). Enforcing Hazardous Waste Rules in India: Strategies and Techniques for Achieving Increased Compliance. Washington, DC, USA: ELI 2014.

[57]    Karthikeyan L, Venkatesan M, Vignesh K. The Management of Hazardous Solid Waste in India: An Overview. Environments 2018; 5(1): 103.
       [http://dx.doi.org/10.3390/environments5090103]

[58]    Mahajan N. Solid waste management in Chennai: Lessons from Exnora. Innov J 2016; 21: 4.

[59]    The Associated Chambers of Commerce of India (ASSOCHAM) and PricewaterhouseCoopers. 2017.http://www.assocham.org/newsdetail.php?id=6350

[60]    Case Study 3. Hazardous Waste Issues in India http://www.eolss.net/sample-chapters/ c09/e1-0--06.pdf

[61]    GoI (Government of India). Twelfth Five Year Plan (2012–2017): Faster, More Inclusive and Sustainable Growth; GoI: New Delhi, India, . 2013.2013.

[62]    Kumar S, Smith SR, Fowler G, *et al.* Challenges and opportunities associated with waste management in India. R Soc Open Sci 2017; 4(3): 160764.
       [http://dx.doi.org/10.1098/rsos.160764] [PMID: 28405362]

# CHAPTER 5

# The E-waste Situation in India and Health Impact on Population

## S.V. A.R. Sastry[*]

*Associate Professor, Department of Chemical Engineering, MVGR College of Engineering (A), Vizianagaram, 535 005, India*

**Abstract:** The chapter provides a brief vision into the perception of E-waste, its production in India, along with the ecological and health issues involved in it. The condition is startling because India produces around two million tonnes of E-waste per year, and practically all of it gets into the natural segment due to the lack of any substitute existing at present. Particularly, cosmopolitan cities like Mumbai, Delhi, and Bangalore are at higher risk of environmental contamination because of E-waste. Personnel in the E-waste dumping segment are ill-protected. They disassemble E-waste, regularly with hands, in awful situations. Nearly 26,000 workers are working in scrap-yards at Delhi itself, where 15 000 to 25 000 tonnes E-waste is moved annually, with PCs alone contributing to 25% of E-waste. Additional E-waste scrap-yards are present in Chennai, Mumbai, Bangalore, Firozabad, and Meerut. The dangerous elements present in E-waste comprise considerable amounts of chromium, cadmium, lead and other fire-resistant plastics. Cathode-ray components and tubes with great lead content are very hazardous for health. Huffing or handling such materials and being in contact with these on a daily basis, may harm the nervous system, brain, kidneys, lungs and also the reproductive system. Functioning in poorly ventilated and bounded areas without practical knowledge and masks leads to contact with hazardous chemicals. The absence of experience made the people endanger their environment and health. There is a dire need for progress in E-waste management encompassing technical development, operative strategy, protecting procedures for the employees involved in E-waste dumping.

**Keywords:** Cathode-ray components, Contamination, Cosmopolitan, Dangerous, Disassemble, Dumping, E-waste, Ecology, Elements, Environment, Fire-resistant plastics, Harm, Hazardous, Health impact, India, Nature, Population, Production, Scrap-yards, Workers.

---

[*] **Corresponding author S.V. A.R. Sastry:** Associate　Professor, Department of Chemical Engineering, MVGR College of Engineering (A), Vizianagaram, 535 005, India; E-mail: svarsastry@yahoo.com

**Gabriella Marfe & Carla Di Stefano (Eds.)**

## INTRODUCTION

A wide range of wastes is imported as well as produced in India, namely E-waste, municipal waste, and dangerous industrial waste. The amount of wastes produced has symbolised a consistently expanding danger for the general well-being of the population. More than eighty-eight fundamentally dirtied manufacturing zones have been recognised by the Central Pollution Control Board (CPCB). Pollutants from these zones taint water streams and dirty groundwater in numerous spots. Studies have additionally demonstrated that crop yields are polluted through modern effluents, yet the size of this effect is still not recognised [1].

To the extent E-wastes is concerned, it has developed as one of the fastest developing waste streams in the current scenario. PCs and hardware gears are planned without giving proper consideration to the downstream impacts and the simplicity of reusing. For whatever length of time, the electronic items contain a combination of dangerous synthetic concoctions and are planned without reusing approaches, they would be representing a risk to the general well-being of the population. As electronic items are at present comprised, E-waste reusing activities in any nation will create contaminating deposits and releases [2]. India has more than one million out-dated PCs with manufacturers, including around 1 051 tonnes of electronic pieces each year. India generates around two million tonnes of E-waste every year. E-wastes now marks more than 71 per cent of landfill. When the developing countries like India begin fixing and authorising severer enactment on trans-limit developments of E-wastes, developed countries consider it difficult to stay away from the concern of reusing and transfer. Nonetheless, exchange bans will turn out to be progressively immaterial in taking care of the issue of E-waste. Also, it is estimated that by 2030, the developing countries would produce double the amount of E-wastes than the developed countries.

Thinking about the future situation, it is rudimentary that the safe managing of E-waste is carried out in a sorted out way. In New Delhi, after the Mayapuri radiation release episode in April, 2010 the administration had issued rules and warnings to all the heads of emergency clinics, medicinal focuses, demonstrative focuses and restorative labs utilising radioactive hardware and consumables for their sheltered transfer. Unexpectedly, underneath the Atomic Energy Regulatory Board (AERB) mandates, the standards recommending Nitti gritty rules with respect to the medicinal introduction, potential presentation, individual observing, quality control, and radiological well-being of officials exist. It is projected that all over India, about five hundred thousand people work in the recycling of E-waste. These occurrences feature the need for an immediate disaster convention, potent guidelines, and inspection mechanism to guarantee that the standards are met. It additionally requires the administrative framework to take into consideration the insurance of labourers. There must be appropriate rights for residents to take a lawful plan of action for harms caused to their well-being, condition and property.

## INTERNATIONAL TRADE OF HAZARDOUS WASTES

Amongst the comprehensive understandings, Basel Convention has the most far-reaching global ecological concurrence on hazardous and diverse wastes. It was embraced during 1989 and came in power during 1992 to secure the well-being of humans and the earth against the antagonistic impacts arising because of the development and transfer of unsafe E-wastes [3]. Initially, it didn't specify E-wastes, yet afterward, it tended to the concerns of E-wastes alongside part of the bargain at Basel Agreement in late 2006. As of now, E-wastes from cell phones consisting of Polychlorinated Biphenyls (PCBs) are utilised in manufacturing units as heat transfer liquids in capacitors and transformers. A significant number of global E-wastes that are sent out in this manner, are in opposition to the Basel Convention.

### Increasing Unlawful E-Waste Trades

As global exchange streams extend and great residential checks elevate the expenses of risky wastes transfer in developed countries, chances and motivating forces for unlawful dealing of wastes will keep on growing.

Numerous examinations have affirmed and uncovered the peril presented by numerous wastes, their harmfulness, cancer-causing nature, and different attributes hurtful to the human well-being and condition. This mindfulness has been the premise of global activity, prompting the fixing of laws and guidelines. This has thus set off expansion at the expense of risky waste transfer through more secure methods convincing numerous countries to scan for all the more financially reasonable methods for arranging E-waste. Therefore, many developed countries, which can go around the national enactments, send out risky wastes including E-wastes to developing countries, which neither are having information of hazardous nature, nor the ability to dispose of the E-wastes securely. For example, in the US, a PC recycler is used for filtering the approaching E-waste materials before selling them to particular representatives. Apart from this, the E-wastes are auctioned off in mass with no partition at all. E-waste facilitating is a forceful and aggressive business, and purchasers for a wide range of E-waste in the Asian market are frequently accessible [4].

## Key Elements in International E-Waste Trade

Similar to other waste exchange, E-wastes cost to developing countries is administered by global financial matters. Again, sending out E-wastes is worthwhile for exporter countries than reusing or arranging it within the country. For example, E-wastes dealers in the USA or Europe need to give 21 US $ to reuse a PC securely in their countries, whereas they can trade it at a large portion of the expense to the casual brokers in developing countries. Still, about 175 US $ is spent to reuse a huge amount of trash after isolation [5, 11].

The US developed multiple times progressively dangerous E-wastes in 2003 (266 million tonnes) than in 1976 (58 million tonnes). The expense of overseeing such E-wastes in the country will be huge, relying upon the reactivity and danger of materials. In this manner, it would be increasingly efficient to transport harmful E-wastes to developing countries when the expense is unimportant. Considering its cost-viability, trade is a furtive choice picked by certain organisations in the industrialised countries. The illicit fares are, for the most part, defended as 'philanthropy' or as 'reusing'. Through these strategies, old gadgets discover their way out from industrialised countries to developing countries where they are utilised for a couple of years more [6, 11]. For example, in 2006, about 1.4 million Personal Computers were of the model 486s, out of almost five million present in India. Reusing may drag out the life expectancy of an item. Possibly, it would discover its way into the waste. In this way, developed countries legitimately sidestep the issue of waste transfer; developing countries are left to figure with a definitive issue of waste transfer.

## Waste Exchange as an Essential Part of Electronics Reprocessing

Bringing in waste is not uncertainty but a rewarding economy. The principle objective after the ingress of utilised gadgets, is the restoration of cost-effective metals and components confined in E-wastes. These different materials give valuable crude material feedstock for the assembling of new items. Nonetheless, this exchange has turned into an essential part of the hardware reusing.

Besides, a considerable lot of the business sectors for prepared plastics and other crude materials are outside the US. The interest for CRT glass cullet stays solid and the quantity of glass heaters keeps on being deficient and lacking [7].

## Waste Trading Through Free Trade Agreements

A quietened part of the global waste exchange is that the developed countries are utilising the "Free Trade Agreements (FTAs)" or alleged "Economic Partnership Agreements (EPAs)" to send out their loss to developing countries. Regularly associated with the EPA plans are implicit remuneration arrangements, for example, the Philippines guaranteed admittance to local and tending work showcases in Thailand and Japan [8, 11].

From 2005, Thailand and Japan are officially arranging an FTA which tries to dispense with duties on a phenomenal rundown. E-wastes get build-up from burned municipal wastes, associated enterprises, and emergency clinical wastes. Developed countries are sending out waste towards Southeast Asian countries like the Philippines, Indonesia, and Thailand through prevailing escape clauses that license a few types of waste being sent for reuse. EU and Japan are now arranging a comparative FTA along with India that can bring about gigantic increment in the ingress of E-wastes seriously hindering natural shield processes [9, 11].

Putting the burden on developing countries to ingress E-wastes through respective or facilitated commerce understandings is a reason for genuine worry because it supports reusing of E-wastes. This can likewise supersede the current global and national regulations against the perilous waste ingress. For example, regardless of the global boycott, the UK sent out almost 23 500 MT of E-wastes "unlawfully."

## INGRESS OF HARMFUL E-WASTES IN INDIA

India is a major waste receiving country on the planet. A wide range of wastes is considered as crude materials, which include dangerous and lethal wastes. In 2018, India generated about seven million tonnes of dangerous waste locally and brought in an additional seven million tonnes.

### India's View on Relaxing Import Rules

The global exchange of re-manufactured items has officially cut across 120 billion US $. Similar to other countries, India has also felt the weight from the developed countries to change its ingress principles to enable admittance to its business sectors for their re-made merchandise. It is contended by countries like Japan, Switzerland, and the US, that advancing exchange of re-fabricated merchandise helps both developing and the developed countries by expanding access to a minimal effort, unrivalled quality items while helping strong waste administration and empowering move of innovation and abilities [10].

## Gaps in Regulations

India's EXIM (trade import) arrangement permits the import of used PCs below ten years of age, other than giving PCs accessories as gifts. These PCs can be given to alleged non-business instructive foundations, enrolled magnanimous medical clinics, open libraries and open supported innovative work foundations.

According to the recommended E-wastes rules of 2011, the provision for ingress of used electronic and electrical hardware for use has been removed in India [11, 12].

## MANAGEMENT OF E-WASTE

The issue of electronic and electrical hardware transfer, import, and reuse has turned into the topic of discussion and dialogue amongst the Government associations and the private division fabricators of PCs. The Committee has recommended a progressively pre-emptive job for CPCB by expressing that it "should lead concentrates to create forthcoming projections and contrive steps to inspect the danger."

In India, the Constitution appoints strong waste administration as an essential duty to the Municipalities under the 12th Schedule. Article 243-W engages the State Legislatures to outline enactments about waste administration. A portion of the rules for taking care of city's tough wastes is imperative for the administration of E-wastes. The rules incorporate sorting out; a gathering of wastes; accumulation of wastes from ghettos, inns, eateries, office buildings, and business regions; arranging mindfulness programs for isolation of wastes; receiving reasonable waste handling innovations; and limiting area filling for non-biodegradable idle waste.

Besides, it is seen that with expanding urbanisation, discovering landfill locations would become hard for the regularly expanding volumes of strong wastes. The Government needs to guarantee that in light of a legitimate concern for general well-being, such landfill destinations will be situated in 'remote spots' [13]. The rules needed to give protected support between human settlement and landfill destinations. Strong waste administration is a necessary piece of the current and prospective urban advancement and recharging plans and projects.

In addition to the above-mentioned details, the E-waste comprises many high-valued and rare materials, such as platinum, gold, and cobalt. The process of inappropriate recycling of rejected electronics leads to significant loss of rare and valued raw materials and puts excessive pressure on our inadequate natural resources. The worth of E-waste in the world is about 62.5 billion US $ per year, more than the GDP of many countries on the planet. Examples of E-waste recycling sites are shown in Fig. (**1**).

**Fig. (1).** E-waste recycling sites.

## HEALTH THREATS AND ENVIRONMENTAL CONCERNS

As indicated by the Ministry of Environment and Forests, there are about twenty-eight active Storage, Treatment, and Disposal Facilities (STDFs) for dangerous waste administration in the country. The rising personal satisfaction and high paces of asset utilisation examples have had an inadvertent and antagonistic effect on the earth.

In addition to the weight of administration of dangerous civil waste, the administration of gigantic and developing E-wastes is rising as one of the most significant natural issues of developing countries, particularly India. Around three hundred thousand tonnes of E-waste were created in India in 2018. With the forecast that about one million tonnes of E-wastes will be created, India's situation in this way faces a genuine danger.

The issues related to E-waste are currently being perceived. E-waste is profoundly mind-boggling to deal with because of its structure. It contains various segments containing harmful elements that adversely affect human health if not taken care of appropriately. Frequently, these issues emerge out due to inappropriate reusing and transfer methods [14, 15].

## The Effect of Dangerous Materials on the Environment and Health

Toxins in E-waste are ordinarily packed in-circuit sheets, plastics, batteries, and LCDs (Liquid crystal displays). Most of the electronic goods contain a huge amount of dangerous metals and synthetic concoctions like mercury, which is as of now being abolished in different countries. Mercury is versatile and toxic in any structure - inorganic, natural or basic. It can produce sorrow and self-destructive propensities and cause loss of motion, Alzheimer's ailment, speech and vision weakness, sensitivities, hypospermia, and feebleness. Mercury bio-gathers (develops in life forms) and biomagnifies (climbs the natural pecking order) [16, 17].

E-waste comprises of intricate blends of constituents and parts down to infinitesimal levels. In any case, because E-waste additionally comprises a critical grouping of elements dangerous to human well-being and earth, even a modest quantity of E-wastes entering will present moderately great measure of substantial halogenated substances and metals. Such harmful elements drain into encompassing water, soil and pass into the air during waste management process or while they are left to lie around or discarded in landfills close to it. At some point or another they would antagonistically influence human well-being and biology.

Except if appropriate well-being actions are taken, the dangerous elements can fundamentally influence the soundness of people in the region – who physically categorise and treat waste – by entering their body from skin, respiratory tract, or through the mucous layer of mouth and the stomach related tract [18].

There is no uncertainty that it has been connected to the developing frequency of a few deadly or seriously incapacitating well-being conditions, including disease, respiratory and neurological issue and birth comeuppances. This effect is observed to be more regrettable in developing countries like India where individuals occupied with reusing E-waste are generally in chaotic areas, living in nearness to landfills or dumps of unprocessed E-waste and working with no assurance or protections. Numerous labourers occupied with these reusing activities are poor and ignorant of the related perils. For example, such reusing procedures lead to decay of nearby potable water, bringing serious ailments.

## Dealings with E-Waste

Right now around the globe, the volume of out-dated PCs and other E-wastes incidentally put away for reusing or transfer is developing at a disturbing rate. The age of the gigantic amount of E-waste puts a tremendous danger to any network. This is shown in Table **1** beneath, which demonstrates the approximate measure of waste made by 500 million PCs.

**Table 1. Ingredient *vs.* Weight of 500 million PCs.**

| Ingredient | Weight (in Pounds) |
|------------|--------------------|
| Plastic | 6.33 Billion |
| Lead | 1.59 Billion |
| Cadmium | 3.1 Million |
| Chromium | 1.95 Million |
| Mercury | 0.63 Million |

There are essentially four ways of treating E-wastes. The most widely recognised one is putting away E-wastes in landfills, yet it is loaded with the threats of filtering and leaching. The perilous impacts are more regrettable in more seasoned or less rigorously kept up landfills. In the US, around 70 percent of substantial metals (counting cadmium and mercury) present in landfills originate from electronic goods. Because of its unsafe nature, discarding in landfills has been restricted in the vast majority of the states in the European Union and the US.

Another technique generally utilised is to burn or consume the products concerned, however, this procedure discharges substantial metals, for example, cadmium, lead, and mercury into the environment. Metropolitan incinerators are probably the biggest point hotspots for dioxins in Canadian and US situations and also of substantial metal pollution of air.

Reusing has been the best since it increments the life expectancy of the items. Reusing comprises direct recycled use after the minor changes are carried out for the working hardware like memory redesigns and so on. In any case, they end up as waste in the long run as they have constrained life expectancy.

While reusing gives off an impression of being a sheltered strategy to use or arrange E-wastes, it tends to be a deceptive portrayal of unique works including destroying, copying, trading and so on which are generally unregulated. "Reusing" of risky wastes, even in best conditions, has little advantage as it essentially moves the perils into auxiliary items that in the long run must anyway be discarded.

## CONCLUDING REMARKS

The hypothesis made in the manuscript is thus confirmed with all the required evidences. The following steps need to be taken for effective management of E-waste:

- strict health safeguards and ecological safety laws in India;
- extended Manufacturers Responsibility in handling of the electronic items;
- import of E-waste only under permit;
- manufacturer-customer-government teamwork;
- conducting Awareness Programmes;
- selecting harmless green IT;
- checking the Compliance of Rules;
- operational monitoring mechanism consolidated by technical capability and manpower;
- source reduction of waste;
- identifying the disorganised segment in India.

## CONSENT FOR PUBLICATION

Not applicable.

## CONFLICT OF INTEREST

The authors confirm that the contents of this chapter have no conflict of interest.

## ACKNOWLEDGEMENTS

Declare none.

## REFERENCES

[1]     Ramesh S, Joseph K. Electronic waste generation and management in an Indian city. J Indian Assoc Environ Manag 2006; 33(2): 100-5.

[2]     Jim P, Ted S. Exporting Harm: The High-Tech Thrashing of Asia. Basel Action Network 2002, pp. 3 - 12

[3]    Solomon UU. A detailed look at the three disciplines, environmental ethics, law and education to determine which plays the most critical role in environmental enhancement and protection. Environ Dev Sustain 2010; 12(6): 1069-80.
[http://dx.doi.org/10.1007/s10668-010-9242-z]

[4]    Sinha Khetriwal D, Kraeuchi P, Schwaninger M. A comparison of electronic waste recycling in Switzerland and in India. Environ Impact Assess Rev 2005; 25(5): 492-504.
[http://dx.doi.org/10.1016/j.eiar.2005.04.006]

[5]    Cucchiella F, D'Adamo I, Koh SL, Rosa P. Recycling of WEEEs: An economic assessment of present and future e-waste streams. Renew Sustain Energy Rev 2015; 51: 263-72.
[http://dx.doi.org/10.1016/j.rser.2015.06.010]

[6]    Sharma HD, Gupta AD. The objectives of waste management in India: a futures inquiry. Technol Forecast Soc Change 1995; 48(3): 285-309.
[http://dx.doi.org/10.1016/0040-1625(94)00066-6]

[7]    Dwivedy M, Mittal RK. Estimation of future outflows of e-waste in India. Waste Manag 2010; 30(3): 483-91.
[http://dx.doi.org/10.1016/j.wasman.2009.09.024] [PMID: 19857950]

[8]    Ravi V. Analysis of interactions among barriers of eco-efficiency in electronics packaging industry. J Clean Prod 2015; 101: 16-25.
[http://dx.doi.org/10.1016/j.jclepro.2015.04.002]

[9]    Garlapati VK. E-waste in India and developed countries: Management, recycling, business and biotechnological initiatives. Renew Sustain Energy Rev 2016; 54: 874-81.
[http://dx.doi.org/10.1016/j.rser.2015.10.106]

[10]   Sthiannopkao S, Wong MH. Handling e-waste in developed and developing countries: initiatives, practices, and consequences. Sci Total Environ 2013; 463-464: 1147-53.
[http://dx.doi.org/10.1016/j.scitotenv.2012.06.088] [PMID: 22858354]

[11]   Dwivedy M, Mittal RK. Willingness of residents to participate in e-waste recycling in India. Environ Dev 2013; 6: 48-68.
[http://dx.doi.org/10.1016/j.envdev.2013.03.001]

[12]   Sastry SVAR. Murthy ChVR. Management of E-waste in the Present Scenario. IACSIT Int J Eng Technol 2012; 4(5): 543-7.
[http://dx.doi.org/10.7763/IJET.2012.V4.428]

[13]   Khoshoo TN, Moolakkattu John S. Mahatma Gandhi and the Environment. TERI Press 2010; pp. 20-30.

[14]   Sastry SVAR. E-waste management in the Indian scenario. Environmental Health Conference. Salvodar, Brazil 2011; pp. 166-69.

[15]   Rakesh J. E-Waste: Implications Regulations and Management in India and Current Global Best Practices. TERI Press 2008; pp. 33-41.

[16]   Sunitha Narain. IT's underbelly. Down to Earth 2010; 19(1): 11-20.

[17]   Kumar Asha Krishna. Importing danger. Frontline 2003; 20(25): 25-30.

[18]   Dutta SK, Upadhyay VP, Sridharan U. Environmental management of industrial hazardous wastes in India. J Environ Sci Eng 2006; 48(2): 143-50.
[PMID: 17913193]

CHAPTER 6

# Hazardous Waste Management and Geological Aspects in Campania (A Case Study)

**Caputo Gaetano**[*]

*I.C. "F. Palizzi", Piazza Dante, 80026 Casoria Naples, Italy*

**Abstract:** The environmental situation is very complex, in Italy, and in particular, in some areas (Naples and Caserta provinces) of the Campania region that have experienced numerous problems correlated with hazardous waste management. In particular, an area of Naples province has been referred to "Land of Fire" (LoF) (or Terra dei fuochi-TdF) for the open burning of uncollected trash, including chemical and other potentially hazardous waste. Different academic publications and the national press have reported this dramatic situation. Findings from several articles suggest that the toxic wastes dumping is destroying this land and in addition, it is seriously damaging the health of the local population. Moreover, the high anthropization of these provinces in association with the simultaneous presence and interaction of extremely active volcanic, tectonic and morpho-dynamic phenomena increases the environmental risk in this territory. In this scenario, the Ministry of Italian Health commissioned epidemiological and geological studies to evaluate both contaminations of soil due to illegal dumps and the health risks on the population in Campania. This chapter aims to examine the epidemiological data considering a geochemical/environmental perspective to better understand the correlation between the incidence of different diseases (such as some cancers type) and the distribution patterns of contaminants.

**Keywords:** Anthropic pollution, Geology of Campania, Geological risks, Waste.

## INTRODUCTION

Recent scientific studies have focused on the relationship between geology factors and human health to better evaluate the geological-environmental effects associated with the distribution of pathologies in living beings. Indeed, some reports observed that particular diseases were more widespread in some geographical areas than in others. In particular, heavily urbanised areas are under constant environmental stress due to human overcrowding and poor waste management. The acquisition of fundamental scientific knowledge of geological,

[*] **Corresponding author Caputo Gaetano**: I.C. "F. Palizzi" Piazza Dante, 80026 *Casoria* Naples, Italy;
Tel /Fax: +390817580785; E-mail: caputo.geo@gmail.com

environmental, chemical and medical nature is essential to understand the cause-and-effect relationship between environmental factors and health problems. Therefore, the in-depth knowledge of all chemical-physical-biological components of a territory and the assessment of possible variations due to pollution are basic components to monitor the territorial resources. Moreover, the characterization of these components can allow the development of a new conception of industrial and anthropic activities, taking into account the serious consequences on the environment and the organisms.The adverse impact of the environment on human health and the development of diseases including cancer [1 - 4] are shown by the combination of epidemiological studies and environmental geochemical investigations. In particular, environmental geochemical studies can determine the sources of pollutants, their behaviour and other characteristics. For this reason, environmental geochemical research can be used to demonstrate the risk for human health due to pollution. For example, numerous studies concluded that long-term exposure to high levels of substances (containing As, Ni, Cr, Cu, Cd) can cause different kinds of diseases such as cancer [5 - 7]. Manganese exposure can induce Parkinson's disease, while elevated levels of Arsenic (As) in waters and sediments can cause skin, bladder and lung cancer. The International Agency for Research on Cancer (IARC) has classified arsenic and arsenic compounds as carcinogenic to humansand has also stated that arsenic in drinking-water is carcinogenic to humans. Other adverse health effects that may be associated with long-term ingestion of inorganic arsenic include developmental effects, diabetes, pulmonary disease, and cardiovascular disease. Arsenic is also associated with adverse pregnancy outcomes and infant mortality, with impacts on child health [8], and exposure in utero and in early childhood is linked to increased mortality in young adults due to multiple cancers, lung disease, heart attacks, and kidney failure [9]. Numerous studies have demonstrated the negative impacts of arsenic exposure on cognitive development, intelligence, and memory [10 - 14]. Based on the data from the geochemical studies, it is possible to monitor the quality of soils and water to avoid that highly toxic heavy metals can enter the food chain.

## NOTES ON GEOLOGY, GEOMORPHOLOGY AND HYDRO-GEOLOGY OF CAMPANIA (ITALY)

Campania region is located in Southern Italy, between the Tyrrhenian Sea to the West and the Apennine mountain chain to the East. There are five provinces of the region: Napoli, Salerno, Caserta, Avellino and Benevento. The region has a population density of 429 inhabitants/km$^2$, an (ISTAT, 2015), and more than 50% of the population concentrated in the Naples metropolitan area (Figs. **1-3**).

**Fig. (1).**  Map of Italy (division: regions) (https://www.maps4office.com/nuts-region-map-italy/).

**Fig. (2).**  Campania Region, southern Italy. (https://italiots.files.wordpress.com/2011/01/italia.jpg).

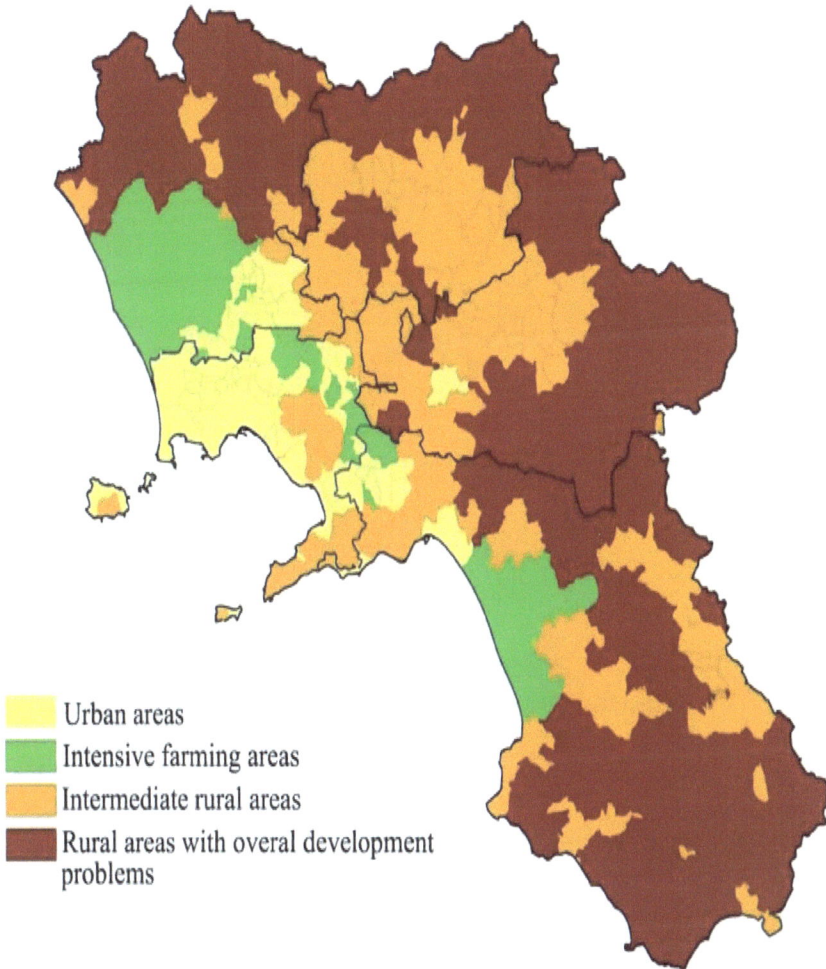

**Fig. (3).** Territorial classification of the 2014-2020 PSR of Campania. (All. 1 at the PSR of the Campania Region – Classification of the rural areas of Campania for the 2014-2020) https://opencoesione.gov.it/media/uploads/documenti/po_approvati/psr/psr_campania.pdf.

The geological structure of Campania is closely linked to the events that have generated the structural picture of the Italian peninsula. The main Italian geological structures are represented by four structural elements: 1) Tyrrhenian area, characterized by a continental crust and, in some areas, by an oceanic crust

2) Apennine chain, consisting of covering blankets and reservoir filling deposits 3) Apennine foredeep, consisting of Plio-quaternary sediments partly buried under the Apennine aquifers 4) the foreland, consisting of a Mesozoic carbonate sequence, set on a continental crust [15]. Campania region presents a hilly and mountainous morphology occupied by the Apennine chain and a coastal sector to the west. Furthermore, it is characterized by the presence of wide structural depressions occupied by alluvial plains (Campania Plain and Sele Plain). The Campania region also consists of four volcanic centres: Roccamonfina in the Caserta area at the border between Lazio and Campania, Vesuvius and Phlegraean Fields close to Naples, and the volcanic complex, the island of Ischia [16]. In particular, from the geological point of view the Campania Plan (CP) area represents a graben originated between the upper part of the upper pliocene and the lower Pleistocene [17]. Throughout the quaternary, the plain has been subjected to a marked burial [18] and such tectonic evolution has led to the formation of direct faults, which border the plain, with NW-direction Apennine S-E and anti-Apennines E-W, whose total vertical discards are around 5000 meters. The genesis of the volcanic phenomena of the Phlegraean area, such as Roccamonfina and the Somma–Vesuvius are linked to such recent structures.The large area has been affected by the deposition of a powerful accumulation of marine deposits of transition environment, by river clastic deposits and volcanic deposits, in both continental and marine facies, from lakes and marshes, as shown by numerous deepgeophysicalsurveys [19]. From the hydrogeological point of view, the geo-lithological and structural characteristics, can affect the deep and superficial water circulation at both on a regional and local scale. Due to the wide grain size range, typical of inconsistent deposits, there are numerous variations in permeability both horizontally and vertically level. These variations result in relatively complex water circulation due to the presence of overlapping aquifers. The most superficial aquifers are fed mainly by direct zenith contributions [20], while the deep aquifer, more powerful than the previous one, is affected by water flows at a regional scale, fed by carbonate reliefs that delimit the CP. Some authors, on the basis of data obtained from phreatic altimetric stationsand from numerous direct observations in the wells, have drawn up maps of the water tables. These tables show that the main aquifer of the plain is directly fed by the meteoric water and additionally, it receives the water from Roccamonfina and Massico mountains, from the north-eastern edge of Casertano and Nola Mountains, from the south-eastern edge of the Sum Vesuvius volcanic complex and at the end, from the southern edge of the Phlegraean hills. In this chapter, we focus our attention on the loose sediments of the alluvial plains where high-volume toxic waste materials have been buried. Large quantities of pollutants from different sources have been discharged into surface and groundwater and have contributed significantly to the quality degradation of groundwater. Finally,

many tonnes of illegally abandoned waste have been burnt, releasing toxic gases into the atmosphere close to urban centres.

## GEOLOGICAL RISKS AND POLLUTION FACTORS FROM NATURAL AND ANTHROPOGENIC SOURCES IN CAMPANIA

The Campania is characterized by the presence of geological, volcanic, tectonic andmorpho-dynamic phenomenaextremely active, which results in various types of risk: volcanic, hydrogeological and seismic activity. The geological conditions, the morpho-dynamic activities and the extensive anthropization of vast regional sectors have made the territory of Campania subject to hydrogeological risk with important consequences on numerous urban centres such as landslides, accelerated erosion, flooding, swelling and shore erosion. In this area, the seismic-triggered landslides are particularly active in the Apennine areas following earthquakes. Furthermore, the Campania region, with the two active volcanoes of Vesuvio and Phlegraean Fields, as well as the volcanic island of Ischia, is the highest volcanic risk region in Italy and indeed one of the highest in the world. Such volcanic system presents different characteristics and activities, with destructive phenomena (pyroclastic fall, base surge, pyroclastic flow, lava flows, lahars). The volcanism of the Phlegraean Fields is characterized by an almost exclusively explosive activity. Such an area is also affected by bradyseism (slow vertical ground movement), related to the resurgence of the central part of the caldera, and by intense fumarolic activity. Naturally, in the Phlegraean and Vesuvian areas, there is a high volcanic risk, considering the zoning of the territory in accordance with the expected danger, defined in the framework of the National Emergency Plans for these areas. The main objective of this chapter is to highlight, if any, an association between exposure to certain specific geological conditions of the territory of Campania with natural sources of pollution, anthropic sources and human health. It is important to characterize an area through a geological in-depth analysis to evaluate elements concentrations that are present into the soil, air and water. In turn, this analysis, depending on composition, may result in adverse health effects in humans, animals and/or plants. Health issues related to a region's geology are visible in both humans and animals on almost every continent. According to some reports released by specialized international agencies such as the United Nations Environmental Program (UNEP), Greenpeace and other international environmentalist organizations, a large quantity of highly toxic wastes derived from Italian and foreign companies are illegally disposed of in Campania. Such illegal dumping of the hazardous wastes in Campania has become a rampant phenomenon after 1990 without rising great social, environmental or political alarm. Furthermore, this hazardous waste had been mainly disposed of in illegal dumps without any concern for environmental safety. During these years, this waste was buried or burned in the areas located north/north-east of the cities of Naples and Caserta. The eco-system of these areas was polluted by such massive environmental devastation by contaminating water resources and agricultural fields. As a direct consequence, the agriculture and breeding of different animals such as sheep and buffalo began to be challenged, since many animals such as sheep and buffalo started to get ill. These animals started to get ill because of the high levels of dioxin, and other dangerous elements such as polychlorobiphenyls. For this reason, different geochemical characterization studies have been carried out in Campania to define the actual degree of the environmental contamination, the sources that have caused it, and to improve the knowledge of the main pathways followed by contaminants to diffuse throughout the different environmental media, for example, all illegal practices of toxic waste (such as combustion and their burial in some agricultural lands) caused a strong crisis of the agriculture and food sector in the Campania region. A project of monitoring of plant foods in the entire region has been promoted in order to evaluate contamination levels of lead and cadmium. In a study, the content of lead and cadmium was estimated in 750 vegetable samples by using the atomic absorption spectrophotometry after microwave mineralization. All samples had low levels of such substances, except for two samples of tomatoes presented an excess of cadmium, and one sample of valerian with an excess of lead [20, 21]. In another study, the authors considered the distribution, inventory, and potential risk of organochlorine pesticides (OCPs) (including Hexachlorocyclohexanes (HCHs) and Dichlorodiphenyltrichloro-ethanes (DDTs), and their association with soil properties and anthropogenic factors in soils of the Campanian Plain. They found that the total concentrations of HCHs and DDTs ranged from 0.03 to 17.3 ng/g (geometric mean: GM = 0.05 ng/g), and 0.08-1231 ng/g (GM = 14.4 ng/g), respectively. Generally, the concentration of OCPs in farmland and orchards was higher when compared with land used for non-agricultural purposes. Furthermore, a significant difference in the concentration of OCPs was observed in the soils across the region: the Acerra-Marigliano conurbation (AMC) and Sarno River Basin (SRB) were considered as severely OCP-contaminated areas. The mass inventory of OCPs in soils of the Campanian Plain is estimated to have a GM of 17.3 metric tons [22]. In another study, the status of POPs (Persistent organic pollutants), including organochlorine pesticides (OCPs), polychlorinated biphenyls PCBs and polycyclic aromatic hydrocarbons PAHs) was characterized in samples of soil, air, and bulk deposition collected in Naples metropolitan area (NMA). The data showed that most of these compounds were present in all the studied environmental matrices, especially in some hotspot areas, such as the Bagnoli Brownfield Site and the infamous "Triangle of the Death", where unwanted ecological risk conditions for PAHs and Endosulfan were determined, respectively. Furthermore, high urban-rural gradients for atmospheric PAHs and PCBs were reported in the Naples metropolitan area (NMA), and the urban areas were considered as the emission source of these contaminants [23]. In another paper, Cuoco *et al.* [24], conducted a study on 538 groundwater wells from the superficial aquifer of the CP, using prospective statistical methods in combination with traditional geochemical techniques. The authors have determined the chemical variables that have been enriched by anthropogenic contamination (*i.e.*, $NO_3$, $SO_4$ and U) using $NO_3$ as a diagnostic variable to detect polluted groundwater. They found that nitrate in the shallow groundwater of the CP was a variable exclusively linked to anthropogenic pollution, while sulphate and uranium were diverted from their natural balance. In particular, sulphate was an enrichment in the solution, which altered the natural hydrogeochemical facies of polluted groundwater. In addition, the same study detected uranium concentrations that were attributed to diffuse pollution by using synthetic fertilizers in agriculture [24]. Moreover, it is important to remember that tons of toxic wastes have been dumped in this alluvial plane (CP) without an impermeable surface at the site in order to avoid the contamination of groundwater flows [25 - 33]. In 2005, 2,507 potentially contaminated sites were recorded by Campania region within the Regional Plan for Remediation of Polluted Sites and they were located in the southern part of Caserta province and the northern part of Naples province [34]. An update in 2008 of these sites brought the number to 3,733 and only 13 of the previously identified sites were remediated [35]. Furthermore, a wide contaminated area of Campania (Giugliano-Resit) was identified in 2010. For this reason, a geological assessment was commissioned in this area to evaluate the pollution caused by six landfills where hazardous waste was mixed with urban waste [36]. The study found that the toxic leachate was penetrated into the underground aquifer, and contamination will peak in 2064 [37]. In addition, such waste has been spilled into agricultural areas and illegally burned, usually overnight, releasing a number of hazardous chemicals, including furans and dioxins, a large family of chlorate compounds with 17 highly toxic molecules, including tetrachlorodibenzo-p-dioxins (TCDD) which were recently classified as carcinogenic in animals and humans by the International Agency for Research on Cancer (IARC) [38, 39]. Another study was conducted on two groups of 25 river buffalo cows from the Caserta and Salerno provinces and the authors, also in this case, found a higher mean value of dioxins (PCDDs, PCDFs) and dl-PCBs along with a higher chromosome fragility in the Caserta samples when compared to those collected in the Salerno area [40]. All these events have led to significant and complex contamination of environmental matrices with a crucial impact on earthy and aquatic ecosystems. Recent studies have indicated that anthropogenic activities such as intensive agriculture, industries, urbanization, tourism, and landfills, are involved in a severe exploitation of natural resources by jeopardising their natural regeneration [41, 42]. Furthermore, these activities cause environmental pollution (such as contamination of air and water, toxic waste, *etc.*) and can interfere with geological processes. Other authors characterized a former cave located in Roccarainola (Naples, Italy) for its eventual destination, such as legal landfill site. A hydro-geochemical survey of the area was conducted on drilling of 14 boreholes and four monitoring wells. In this study, the authors found enormous volumes of composite wastes that were located below the free surface level of the aquifer. In addition, the examined samples of the boreholes contained levels of As, Cd, Cr, Cu, Hg, Pb, Sn, Tl and Zn above the legal limits In particular, Cd, Pb and Zn concentrations were in excess of about 50, 100 and 1,870 times the limit. Furthermore, polycyclic aromatic hydrocarbon levels were extremely high. The aquifer system had very high concentrations of Cd, Cr-tot, Cu, Fe, Mn, Ni, Pb and Zn. Furthermore, hot gases up to 62°C (such as xylene and ethylbenzene) were found [43]. Other studies were carried out to obtain both detailed maps and spatial distribution patterns of potentially toxic elements (PTEs) in different matrices. In particular, one research found an elevated spatial abundance of 15 PTEs in urbanized (Naples and Salerno) areas, highly industrialized (Solofra) area and intensively cultivated areas (Sarno River Basin). In the latter area, the authors detected the increased concentration of elements (including Pb, Sb, Sn and Zn) and elevated soil contamination compositional index (SCCI) values compatible with high anthropogenic contamination [44]. Another study analyzed lead isotope in soils from 7 profiles (1 m depth) and in groundwaters from 8 wells in Domizio-Flegreo Littoral. The results showed that Pb isotopic composition was present in superficial soils in more urbanized areas (Giugliano) [45]. Another study was carried out in the Volturno River's lower basin (Campania, south Italy): this area included the "Litorale Domizio-Flegreo e Agro Aversano" Site of Regional Interest (SRI), due to the presence of several illegal landfills. Specifically, this research analyzed the distribution and concentration of selected potentially toxic elements (Cd, Cu, Pb, and Zn) in soil samples. A detailed survey was carried out by collecting 64 core samples at fixed depths (0–20, 20–40 cm) and examining seven different soil profiles through the identification of 39 genetic horizons. In this regards, PTEs distribution did not show a linear concentration gradient with depth. Conversely, evident fluctuations along profiles were systematically observed. In particular, the raw data sets in core drill samples conformed to the background population (median ± 2MAD) for Cu (12.93–48.33 mg kg$^{-1}$), Pb (1.70–50.10 mg kg$^{-1}$), and Zn (32.13–129.13 mg kg$^{-1}$). Furthermore, Cd was present at a high level compared to the geogenic background concentration (0.43 mg kg$^{-1}$) as well as the threshold values (2 mg kg$^{-1}$) obtained by the Italian legislation. In addition, some unusual PTEs concentrations were inconsistent in comparison with "contaminated sites". In conclusion, the authors suggested that the agricultural soils of Volturno River's lower basin required control and protection from the illegal deposition of waste materials [46].

# ENVIRONMENTAL GEOCHEMICAL STUDIES AND HUMAN HEALTH IMPACT

In Campania, the public health authorities did not recognize the link between the increased rates of some diseases such as cancer and the toxic waste dumping because of the lack of reliable data at a medical level. In 2004, the prestigious scientific journal Lancet Oncology published an article [47] that reported the dramatic situation in these areas called the 'triangle of death' (a triangle between the towns of Nola, Marigliano and Acerra, east of Naples). The study showed for the first time that the cancer rate was high in District 73, the public health code of the area, during the period between 1994 and 2000. It was higher when compared with the national and regional average, and for this reason, it was urgent to investigate to the link between hazardous waste and cancer mortality [46, 47]. Furthermore, the epidemiological SENTIERI study, (project, coordinated by the Istituto Superiore di Sanità from 2007–2010), conducted in 55 municipalities in the "Land of Fires", reported increased incidence, hospitalization and mortality rates in the province of Naples for the stomach, liver, lung, bladder, pancreatic, laryngeal, kidney, breast cancer, and non-Hodgkin's lymphoma, while in the Caserta province, excess of mortality and hospitalization rates were reported for the stomach, liver, lung, bladder, laryngeal cancer, leukemia. Furthermore, an increased high hospitalization rate for myocardial infarction in women was also found in the Caserta province [48]. Moreover, some authors detected high levels of dioxins and PCB in human milk-derived by young women (under 32 years old) living in 45 towns located in Caserta and Naples provinces [49]. The geochemical characterization of the areas of Campania with epidemiological studies can be a new approach to define the degree of environmental contamination, the sources of contamination and the main pathways followed by contaminants in order to better understand the health risks for the population (Fig. **4**). For example, radon exposure provides an example of the relationship between geology and negative human health implications. Such gas, the second cause of lung cancer after smoking (WHO- IARC), is a natural, radioactive gas, which originates from the soil and pollutes indoor air, especially in closed or underground spaces. There is a significant production of radon in the Campania region being a volcanic territory. In the world, the average radon concentration is 40 bq/mc, but in our country, it is higher than 70 bq/mc. In Campania, however, it is three times higher than the world's average: about 120 bq/mc [50] for its geological characteristics. Several studies pointed out the lethal impact of radon on humans' health [51, 52]. Above all, the studies confirmed the association between the strong risk of lung cancer and radon concentrations [53, 54]. Moreover, 9% of deaths from lung cancer are caused by indoor radon and about 2% of all deaths from cancer in Europe [54]. The International Agency for Cancer Research (IARC) and the World Health Organization have classified radon as one of the proven group 1 carcinogens.

Since 1988, several national and international agencies have estimated a significant proportion of cases of lung cancer linked to radon, for example, in Italy, between 1500 and 6000 cases per year (out of a total of around 30000) are recorded which could be due to exposure to radon. Therefore, detection of the areas with higher concentrations in outdoor-indoor radon plays a fundamental role in the prevention of risks to human health [54]. The radon-soil gas derived from the rock fragments emanation or crystals and exhalation from pore spaces in the soil produced high indoor radon levels (in house air Radon concentrations) [55]. Therefore, field studies have confirmed a strong relationship between the geology and Radon-soil gas [56 - 60]. Furthermore, other studies have stressed upon the strong association between geology and human health [61, 62]. The interplay between human health and the geological characteristics of the earth's surface is very complex [63, 64] and, additionally, the relationship between exposure and health outcomes varies significantly among different geological hazards within the environment. For example, the interaction between the development of skin cancers and exposure to water and food supplies contaminated with arsenic is well known [65, 66]. Furthermore, the phenomenon of volcanism is involved in the principle processes that bring metals and other potentially dangerous pollutants to the surface from deep within the earth. Indeed, in Campania, the natural sources of potentially toxic substances are limited in certain areas and correlated directly with the lithological properties of the outcrops present. Specifically, the emission of the metal elements is linked to the presence of volcanic products of Somma-Vesuvius, Roccamonfina and the Phlegraean Fields. Natural concentrations of such elements (such as Fe, As, S, Cd, Cu, Hg, U, Zn, Mo, Pb) are due to magmatic differentiation and subsequent transport by residual aqueous fluids, and for these reasons, they can be present in high concentrations in the rocks where they infiltrate [62]. Heavy metals also derive from anthropogenic activities that cause their large production in different environmental compartments. A preliminary study pointed out the presence of several heavy metals and organic compounds both in the soil and groundwater in contaminated sites in the North of Naples [67]. In another study, the Standardized Incidence Ratio (SIR) of various cancer types in the local population of the Salerno province was correlated with the distribution of some toxic metals in the stream sediments of the whole province. Such a study reported a high rate of lung, liver and prostate cancer in this province, markedly contaminated by heavy metals, above all in the Sarno River plain [68]. In another study, some authors found a strong relationship between geochemical and epidemiological studies in the Campania region. They reported maps that represented the toxic metal concentrations model in association with some potential pathologies. Such a comparison pattern at regional scale has been made considering previous epidemiological research, that showed the possible relationships between anomalous concentrations of some metals and the incidence of some pathologies. According to the previous epidemiological data, this paper showed that some types of cancer were found in some areas that were characterized by a relatively high concentration of heavy metals. These preliminary data suggested a strong association between compromised environment such as the urban and provincial areas of Naples, and negative effects on human health, considering the limitations of epidemiological studies to demonstrate the correlation between pollution and pathologies [69, 70]. Another interesting study evaluated the correlation between overall chemical contamination and the potential risks for human health. The authors used the GIS with the Multi-Criteria Decision Analysis (MCDA) to perform the Spatial Multi-Criteria Decision Analysis (S-MCDA). First of all, the authors reported that a significant part of total emissions came from road traffic. Furthermore, they found a high concentration of potentially Toxic Metals (PTMs) in many agricultural soils. In such soils different elements (As, Be and Tl, and possibly Sn) had a geogenic origin; while the Cd presence was due to the effect of co-precipitation induced by the Fe and Mn in soils of alluvial origin. Other elements (Cu, Pb and Zn) were produced by heavy motor vehicle traffic, industrial settlements (Pb, Zn) and agricultural practices (Cu to vineyards). This hydrogeological study also found the high concentration of F, As, Fe and Mn that were of geogenic origin. In this case, the authors considered the water depth to be the most significant parameter influencing groundwater contamination (mainly by nitrates of agricultural origin). It is possible to determine an Available Water (AW) index map using the soil hydraulic characterization. Such a map, which represented water stored in the soil profile available for plant growth, is useful for obtaining the "suitability map". Furthermore, this study showed a high rate of incidences of stomach, colon-rectum and liver cancer in both genders, while excess rates of lung and bladder cancer only in males. However, this study presented a number of limitations such as the absence of actual exposure data in the population (biomarkers), lack of information on the main confounding factors present in the population (smoking habits, diet, socio-economic status and deprivation), (long) lag times between any relevant exposure and onset of disease, difficulties in assessing any cumulative, potentiating or synergistic effects from multiple chemicals, and others. Future research will be necessary to better characterize the potential association between exposure and health effects [71]. Another study showed the concentration of 53 elements on 3535 topsoil samples, collected across the whole regional territory. The concentrations were determined using aqua regia extraction followed by a combination of ICP-MS (inductively coupled plasma mass spectrometry) and ICP-AES (atomic emission spectrometry) methods. A new approach to assess/rank environmental risk was applied by using geospatial analysis in a GIS (geographic information system) platform. Such methodology was wide accepted by European countries for the preliminary assessment of human health risks at single contaminated sites at a regional scale. Furthermore, the methodology chosen for the risk assessment procedures is the PRAMS (Preliminary Risk Assessment Model for the identification and assessment of problem areas for Soil contamination in Europe). This integrated approach applied here provided a more robust qualitative and quantitative evaluation, highlighting new and vital information on the distribution and patterns of key elements in soils of the Campania region and the implication on health risk for the local population (Fig. 5) [72]. A complete study of Pb isotope was carried out in several areas of Campania Region (Acerra, Giugliano in Campania, Aversa, Qualiano, Caivano, Casoria, Marigliano, Nola, Pomigliano, Marcianise municipalities, Domizio Littoral and Agro Aversano areas). Furthermore, these areas were recently involved in a waste emergency caused by the inadequate management of solid wastes. The different media for sampling (soil, groundwater, vegetables and hair) were collected in selected areas and analyzed for potentially toxic elements (PTEs): mostly Pb, Zn, Cu, As and Be. In the first phase, 1064 surface soils (0-15 cm), 27 groundwater, 12 corn and 24 hair samples have been collected across the whole study area. All the samples were treated and prepared to determine the concentrations of 53 elements, by using the ICP-MS.Pb isotope ratios were also determined to identify the sources of metals. The data showed that soil sampling sites are characterized by the exceeding presence of As, Cd, Co, Cr, Cu, Hg, Pb, Se, and Zn. In detail, 46% of the topsoil residues, 96% of topsoil leachates, 88% of groundwater, 90% of human hair, and 25% of corn samples pointed out that about 50% of the lead in this area can be due to ascribed to anthropogenic activity. In the second phase, 967 soil samples collected in the Domizio Flegreo and Agro Aversano were analyzed with a methodology that combines the ICP-MS (mass spectrometry Inductively Coupled Plasma) and ICP-ES (Emission Spectrometry coupled plasma inductively). The data reported high concentrations of most critical PTE elements Sb, As, Cu, Hg, Pb, Zn [73]. In the third phase, this study tried to evaluate the impact of the incinerator on both the Acerra territory and public health-based epidemiological research works and to establish a record of the actual environmental conditions. A total of 121 samples were collected to analyze potentially harmful elements PHEs: As, Cd, Co, Cr, Cu, Hg, Ni, Pb, Sb, Se, Tl, V, and Zn. Thiry-three 33 samples were collected for the analysis of PAHs. The data showed excess of concentrations of Pb, Zn and V (limits established by the Italian Environmental law D.Lgs. 152/2006) in the most urbanized areas. Furthermore, the agricultural soils close to the urbanized areas had elevated levels of Cu, Co, Cd, Be and Ni; while in the incinerator area, the concentration of Se, Hg, Cu, Cd and S was generally higher than in the rest of the territory. Moreover, the high PAHs distribution level suggested that the practice of agricultural waste burning in this area could be a strong source of pollution. The geochemical distribution patterns show both high and low molecular weight

PAHs in different areas: the metropolitan area of Naples, the Agro Aversano area, and, partly, the Sarno River basin. In accordance with the Italian environmental law (D.Lgs. 152/2006), these areas should be unsuitable for residential use. As a consequence, a preliminary quantitative risk assessment improved by the use of GIS pointed out cancer risk higher than 1x10-5 for the city of Naples and for some other populous areas [74 - 78]. The second part of the research was finalized based on the analysis of risk assessment for the human health in Acerra. Regarding this part of the research, 178 topsoil samples and 10 food samples (corn and chicorium endive) were taken form Acerra. All the samples were analyzed for 15 elements (As, Be, Cd, Co, Cr, Cu, Hg, Ni, Pb, Sb, Se, Sn, Tl, V and Zn) by ICP-MS and ICP-ES after an aqua-regia digestion. Enrichment factors and pollution indexes of the PTEs in the soil were estimated considering their regional background levels. The results showed low contamination for these metals according to the Italian soil quality standards (D.lgs 152/06), and the element Pb was the most important contaminant in this area. In accordance with the latter data, an increase in the baseline cancer incidence could happen in the resident population up to 2 units per 100,000 exposed toddlers and up to 3 units per 100,000 exposed adults over an exposure time of 6 and 60 years, respectively. In this scenario, it is possible to reliably predict an overall increase in gastric and respiratory tract cancer incidences of 23 units for adults and of 1 unit for toddlers, considering the adults were around 750,000 and the toddlers were around 50,000 (2012-Comune di Napoli 2 www.comune.napoli.it › flex › pages › ServeBLOB.php › IDPagina) and they were all likewise exposed to PAHs in the soil for the same time [79]. Besides in the biomonitoring SEBIOREC study (Studio epidemiologico di biomonitoraggio della Regione Campania-Epidemiological study of biomonitoring of the Campania Region), de Felip *et al.*, measured PCDDs, PCDFs, PCBs and heavy metals levels (lead, mercury, arsenic, and cadmium) in 84 serum and blood samples derived from approximately 850 donors living in 14 municipalities of the Caserta and Naples provinces and 6 milk samples from 52 women from the same areas. The researchers observed that dioxin concentrations in human milk samples were strongly age-dependent and positively correlated to the risk area where mothers were living, while biomarker concentrations in serum and blood samples were comparable with their current values in European countries and in Italy [80]. Furthermore, in recent years, several studies reported that exposure to environmental pollution may contribute to a decrease the semen fertility caused by exposure to heavy metals. Some authors observed that participants with low-quality semen (semen volume, sperm concentration, sperm total count, sperm motility, pH) lived in areas with anomalous Pb and Sb concentrations, considering the geochemical distribution of heavy metals in soils of the Napoli metropolitan area [81].

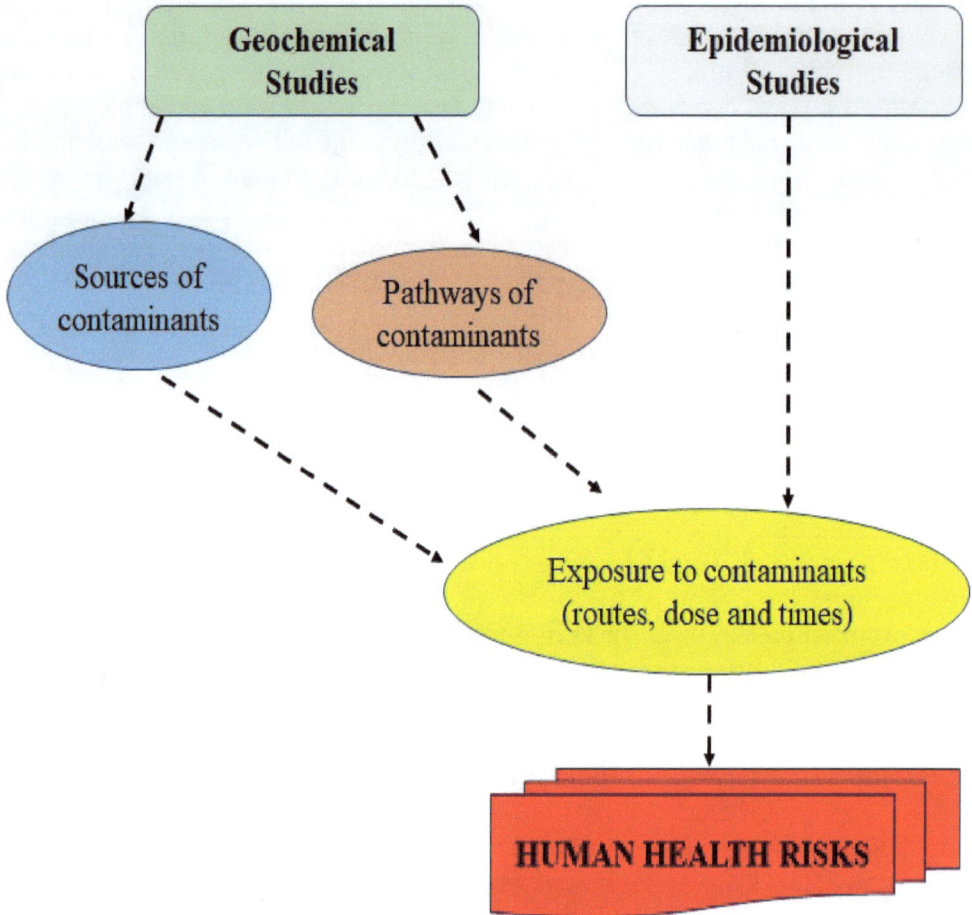

**Fig. (4).** The association framework between geochemical and epidemiological studies to better evaluate the human health risks.

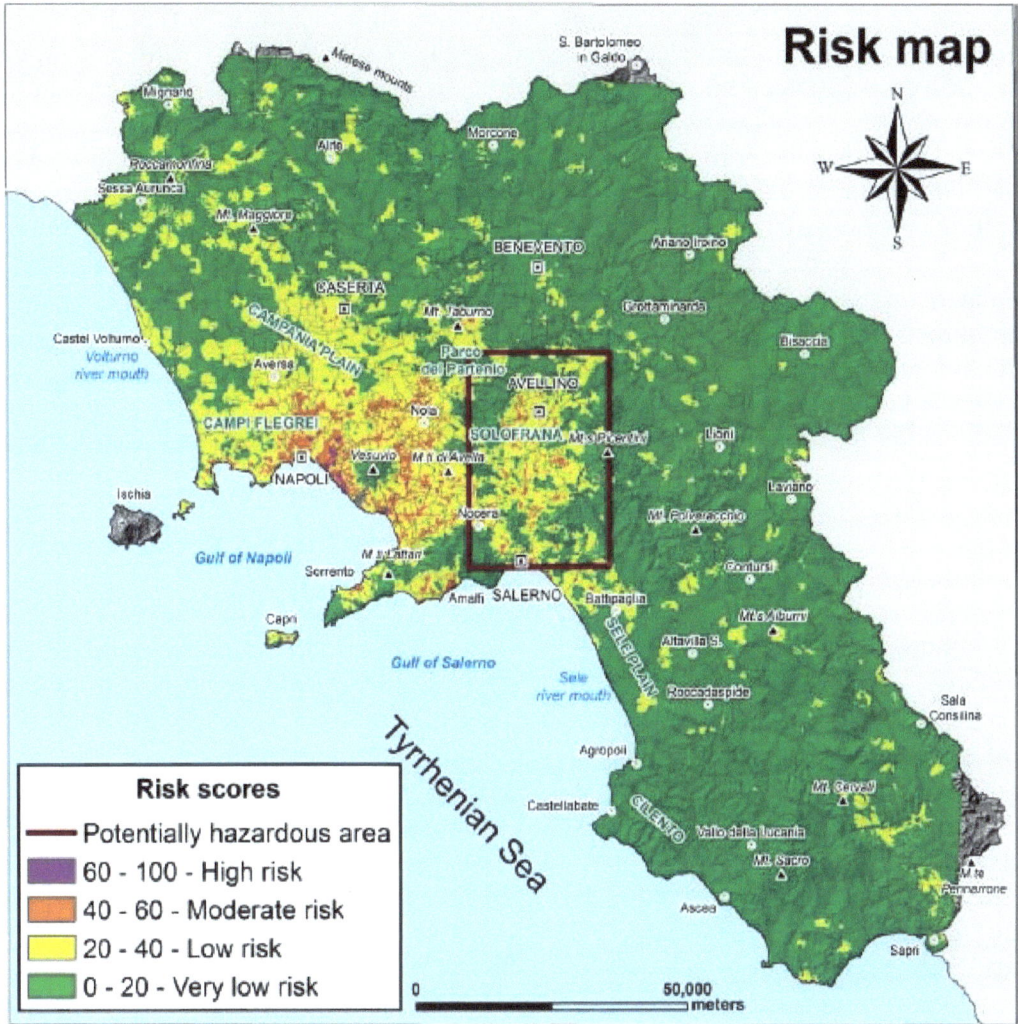

**Fig. (5).** Overall risk map for human health risk assessment in Campania region. (Minolfi G. Albanese S, Lima A Tarvainen T, Fortelli A, De Vivo B. A regional approach to the environmental risk assessment - Human health risk assessment case study in the Campania region. HYPERLINK "https://www.sciencedirect.com/science/journal/03756742"Journal of Geochemical ExplorationHYPERLINK "https://www.sciencedirect.com/science/journal/03756742/184/part/PB" 184, Part B, January 2018, Pages 400-416).

Furthermore, it is important to characterize soil, water, air, and food from a geological point view and to determine the metals and organic compounds' concentration within these media to better carry out the environmental risk analysis. Besides the considered compounds, elevated levels of mercury pose another serious concern in the pollution and disruption of the food chain. According to some estimates, about 80% of the mercury released into the environment comes from natural sources (rock erosion by atmospheric agents and rivers, vaporization from the earth's crust), and the remaining 20% from anthropic origin. According to recent studies, emissions from waste incineration account for 36 tonnes of mercury per year in the European Community. In addition, inorganic mercury in water is converted into methylated mercury in bottom sediments. Methylated mercury compounds are liposoluble and therefore, readily accumulate in living organisms and are concentrated along the food chain. The general population is exposed to methylmercury, mainly through the feeding of fish products that are the dominant source. Furthermore, mercury is extremely toxic to the human body above all for the central and peripheral nervous system, heart, kidneys and immune system. Moreover, multiple sources such as slurry from livestock farms used for decades as an agricultural fertiliser; inorganic pesticides; waste from demolition, construction and hazardous wastes from excavation activities; industrial processing wastes; waste from artisanal processes; commercial waste; waste from service activities; waste from recovery and disposal activities; sludges from drinking water and other water treatments, sewage treatment and fume abatement; sanitary waste; deteriorated and obsolete machinery and equipment; end-of-life motor vehicles, and trailers waste fuel are contributing locally to the pollution of soil, water and air. Soils are enriched or depleted by harmful chemical elements depending on their physical characteristics. In this case, local specific factors such as porosity, the thickness of soil horizons, permeability, presence of groundwater and its direction, effective infiltration, intrinsic characteristics of pollutants, *etc.*, play an important role. The relationship between exposure and health outcomes involves different geological hazards within the environment [82, 83]. In this context, the interdisciplinary approaches are needed to demonstrate the complex interplay between medical geology and human health and to ensure accurate interpretation of environmental exposures and health outcomes [84]. Waste management and disposal are, therefore, crucial for the environment and the health of living organisms. For this reason, the precaution and preventive measures can play a fundamental role in waste management in association with efficiency and transparency to reduce all possible environmental risks and damages to health.

## CONCLUSION

The data analysis points out a mutual relation among geological aspects, anthropic factors, pollution and human health. The earth resources, natural environmental factors and land-use can have a strong influence on human health. Over the past few decades, multiple findings show the impact of agriculture, resource exploitation, urbanization and waste management on human health. Most of the research, focusing on industrial and waste landscapes in developing countries, have demonstrated a strong association between the geology and element intake. All of the abovementioned studies point out the importance of geology and geomorphology in order to characterize and prevent different health problems linked to environmental degradation. For instance, the Italian peninsula, and particularly, the Region of Campania, are characterized by a wide deterioration of environmental conditions. This situation is related to a lack of administration as well as inefficient waste disposal practices. Moreover, it is necessary to highlight the specific inherent risks in the whole area and, particularly, in the Campanian territory. Getting into details, the presence of a high concentration of radon gas in water, soil and atmosphere and a remarkable inflow of chemical elements and toxic substances cannot be ignored. Even though different local governments have made several efforts in order to find a solution to this problem, it still seems an unsolved question. The main threats for the environment are represented by a vicious circle of greenhouse gas emissions, uncontrolled dispersion of plastic materials and several synthetic pollutants. The concern is addressed by the scientific community because of the high concentration of methylmercury into phytoplankton, which represents the basic nourishment for most of the marine fauna. The effects of this situation on the food chain and on human life will be remarkable and easily predictable. Nevertheless, it is difficult to imagine a reduction in industrial productions or increased use of alternative energy sources in spite of the numerous promises made by several nations. It is worth underlining that everyone should take an active part in a worldwide campaign to establish an environmental regulation that aims to limit human pollution and associated damages.

## CONSENT FOR PUBLICATION

Not applicable.

## CONFLICT OF INTEREST

The authors confirm that the contents of this chapter have no conflict of interest.

## ACKNOWLEDGEMENTS

Declare none.

## REFERENCES

[1]     Bennett BG. Exposure of man to environmental arsenic - an exposure commitment assessment. Sci Total Environ 1981; 20(2): 99-107.
[http://dx.doi.org/10.1016/0048-9697(81)90056-5] [PMID: 7302564]

[2]     Spang G. *In vivo* monitoring of Cadmium workers in Cadmium 86 Edited Proceedings. North Carolina: Cadmium Council New York, IL 2RO Research Triangle Park 1988; pp. 162-4.

[3]     NIOSH. (National Institute of Occupational Safety and Health) Criteria for recommended standards: Occupational Exposure to Inorganic nickel. Washington, D.C.: U.S. Department of Health, Education and Welfare 1977; pp. 1-282.

[4]     Varnavas SP, Kalavrouziotis IK, Karaberou G, Apostolopoulou KA, Varnavas PS. Medical geochemical investigations in taking precautionary measures against diseases. protection of human health. Glob NEST J 2012; 14: 505-15.

[5]     EPA, U.S. Environmental Protection Agency. What You Need to Know about Wood Pressure Treated with Chromate Copper Arsenate. CCA 2002.

[6]     Merian E, Ed. Metals and Their Compounds in the Environment. Weinheim: VCH 1991.

[7]     Nriagu JO, Ed. Nickel in the Environment. New York: Wiley 1984; pp. 1-833.

[8]     Quansah R, Armah FA, Essumang DK, *et al.* Association of arsenic with adverse pregnancy outcomes/infant mortality: a systematic review and meta-analysis. Environ Health Perspect 2015; 123(5): 412-21.
        [http://dx.doi.org/10.1289/ehp.1307894] [PMID: 25626053]

[9]     Farzan SF, Karagas MR, Chen Y. *In utero* and early life arsenic exposure in relation to long-term health and disease. Toxicol Appl Pharmacol 2013; 272(2): 384-90.
        [http://dx.doi.org/10.1016/j.taap.2013.06.030] [PMID: 23859881]

[10]    Tolins M, Ruchirawat M, Landrigan P. The developmental neurotoxicity of arsenic: cognitive and behavioral consequences of early life exposure. Ann Glob Health 2014; 80(4): 303-14.
        [http://dx.doi.org/10.1016/j.aogh.2014.09.005] [PMID: 25459332]

[11]    Ravenscroft P, Brammer H, Richards K. Arsenic Pollution: A Global Synthesis. Wiley-Blackwell 2009.
        [http://dx.doi.org/10.1002/9781444308785]

[12]    BBS/UNICEF.. Dhaka: Bangladesh Bureau of Statistics/UNICEF Multiple Indicator Cluster Survey 2012-13 Final Report 2015.

[13]    Argos M, Kalra T, Rathouz PJ, *et al.* Arsenic exposure from drinking water, and all-cause and chronic-disease mortalities in Bangladesh (HEALS): a prospective cohort study. Lancet 2010; 376(9737): 252-8.
        [http://dx.doi.org/10.1016/S0140-6736(10)60481-3] [PMID: 20646756]

[14]    Flanagan SV, Johnston RB, Zheng Y. Arsenic in tube well water in Bangladesh: health and economic impacts and implications for arsenic mitigation. Bull World Health Organ 2012; 90(11): 839-46.
        [http://dx.doi.org/10.2471/BLT.11.101253] [PMID: 23226896]

[15]    Scandone P, Patacca E, Meletti C, Bellatalla M, Perilli N, Santini U. Struttura geologica, evoluzione cinematica e schema sismotettonico della penisola italiana. Atti del Convegno Annuale del Gruppo Nazionale per la Difesa dai Terremoti. 1990; 1: pp. 119-35.

[16]    L'ambiente geologico della Campania, a cura di Antonio Vallario CUEN srl Napoli 2002.

[17]    Brancaccio L, Cinque A, Romano P, Rosskopf C, Santangelo N. L'evoluzione delle pianure costiere della Campania: geomorfologia e neotettonica. Mem Soc Geogr It 1995; 53: 313-36.

[18]    Ortolani F, Aprile F. Nuovi dati sulla struttura profonda della Piana Campana a S-E del fiume Volturno. Boll Soc Geol Ital 1978; 97.

[19]    AA W. La pianura Campana e il suo tufo 1968; 7-11. ottobre

[20]    Celico P. Idrogeologia dei massicci carbonatici, delle piane quaternarie e delle aree vulcaniche dell'Italia centro-meridionale (Marche e Lazio meridionali, Molise e Campania). Quaderno Cassa del Mezzogiorno 1983; 4(2): 172-7.

[21]    Esposito M, Picazio G, Serpe P, Lambiase S, Cerino P. Content of cadmium and lead in vegetables and fruits grown in the campania region of Italy. J Food Prot 2015; 78(9): 1760-5.

[http://dx.doi.org/10.4315/0362-028X.JFP-15-072] [PMID: 26319733]

[22]  Qu C, Albanese S, Chen W, *et al.* The status of organochlorine pesticide contamination in the soils of the Campanian Plain, southern Italy, and correlations with soil properties and cancer risk. Environ Pollut 2016; 216: 500-11.
[http://dx.doi.org/10.1016/j.envpol.2016.05.089] [PMID: 27376995]

[23]  Qu C, Albanese S, Lima A, *et al.* The occurrence of OCPs, PCBs, and PAHs in the soil, air, and bulk deposition of the Naples metropolitan area, southern Italy: Implications for sources and environmental processes. Environ Int 2019; 124: 89-97.
[http://dx.doi.org/10.1016/j.envint.2018.12.031] [PMID: 30640133]

[24]  Cuoco E, Darrah TH, Buono G, *et al.* Inorganic contaminants from diffuse pollution in shallow groundwater of the Campanian Plain (Southern Italy). Implications for geochemical survey. Environ Monit Assess 2015; 187(2): 46.
[http://dx.doi.org/10.1007/s10661-015-4307-y] [PMID: 25638062]

[25]  Legambiente        RS.        Rifiuti        Spa,        disponibile        sul        sito 2012.http://www.legambiente.it/sites/default/files/docs/dossier_rifiuti_spa_rivisto_0.pdf

[26]  Regione Campania (a cura di Commissario di Governo per l'emergenza bonifiche e tutela delle acque nella Regione Campania) Aggiornamento Piano Regionale di Bonifica dei siti inquinati della Regione Campania, Napoli 2011.

[27]  Campania R. Autorità di Bacino Nord Occidentale. Napoli: Piano Stralcio per la Tutela del Suolo e delle Risorse Idriche 2010.

[28]  Zanfi F. Città Latenti Un progetto per l'Italia abusiva. Roma: Bruno Mondadori 2008.

[29]  Bonomi L. Bonifica dei siti inquinati - caratterizzazione e tecnologie di risanamento. Milano: McGraw Hill 2005.

[30]  Campania R. Piano di Tutela delle Acque. Adeguamento al DLgs 152/2006 e smi, Convenzione Regione Campania - Sogesid SpA Rep n13360 del 26 marzo 2003.

[31]  Campania R. a cura di Commissario di Governo per l'emergenza bonifiche e tutela delle acque nella Regione Campania, ARPAC) Piano Regionale di Bonifica dei siti inquinati della Regione Campania, Napoli. 2005.

[32]  Unichim. Suoli e falde contaminati Campionamento e analisi, Manuale 196/2004, Parte- II, Milano, Unichim 2004.

[33]  Viparelli M. Le acque sotterranee ad oriente di Napoli. Napoli: Giannini 1978.

[34]  ARPAC. Piano regionale di bonifica dei siti inquinati della regione Campania 2005. http://www.sito.regione.campania.it/burc/pdf05/burcsp09_09_05/pianoregionale_bonifica.pdf

[35]  ARPAC. Relazione sullo stato dell'ambiente in Campania 2009. www.arpacampania.it/ dettaglio-pubblicazione/-/asset_publisher/    m32W/content/relazione-sullo-stato-dell%E2%80%99ambient--in-campania-2009-2009?redirect=https%3A%2F%2            Fwww.arpacampania.it%2Fdettaglio-pubblicazione%3Fp_p_id%3D101_INSTANCE_m32W%26p_p_lifecycle%3D0%26p_p_state%3Dno rmal%26p_p_mode%3Dview%26p_p_col_id%3Dcolumn-1%26p_p_col_pos%3D1%26p_p_col_count%3D2

[36]  Balestri G. Consulenza Tecnica nei luoghi di cui al decreto di sequestro probatorio del 17/07/2008 e segg nelle località Masseria del Pozzo, Schiavi e San Giuseppiello in Giugliano in Campania; terreni SP Trentola-Ischitella in Trentola e in Torre di Pacifico in Lusciano e siti non sequestrati in Castel Volturno o oggetto di precedenti sequestri in località Scafarea (Giugliano). Firenze: Procura della Repubblica 2010.

[37]  De Rosa SP. Reclaiming Territory from Below: Grassroots Environmentalism and Waste Conflicts in Campania. Lund, Italy: Lund University 2017.

[38]  Steenland K, Bertazzi P, Baccarelli A, Kogevinas M. Dioxin revisited: Developments since the 1997

IARC classification of dioxin as a human carcinogen. Environ Health Perspect 2004; 112(13): 1265-8.
[http://dx.doi.org/10.1289/ehp.7219] [PMID: 15345337]

[39]  IARC.  World  Cancer  Report  2014.  Lyon:  WHO  2014.
https://www.who.int/cancer/publications/WRC_2014/en/

[40]  Genualdo V, Perucatti A, Iannuzzi A, *et al.* Chromosome fragility in river buffalo cows exposed to dioxins. J Appl Genet 2012; 53(2): 221-6.
[http://dx.doi.org/10.1007/s13353-012-0092-2] [PMID: 22415351]

[41]  Slingenberg A, Leon B, van der Windt H, K, R. L, E, Kerry. T. Study on understanding the causes of biodiversity loss and the policy assessment framework. Brussels: European Commission 2009.

[42]  Commission Staff Working Document. The Eu Environmental Implementation Review 2019. Country Report – Spainhttps://ec.europa.eu/environment/eir/pdf/report_es_en.pdf

[43]  Ferrara L, Iannace M, Patelli AM, Arienzo M. Geochemical survey of an illegal waste disposal site under a waste emergency scenario (Northwest Naples, Italy). Environ Monit Assess 2013; 185(3): 2671-82.
[http://dx.doi.org/10.1007/s10661-012-2738-2] [PMID: 22766923]

[44]  Petrik A, Thiombane M, Cima A, Albanese S, Busher JT, De Vivo B. Soil contamination compositional index: A new approach to quantify contamination demonstrated by assessing compositional source patterns of potentially toxic elements in the Campania Region (Italy). Appl Geochem 2018; 96: 264-7.
[http://dx.doi.org/10.1016/j.apgeochem.2018.07.014]

[45]  Grezzi G, Ayuso RA, De Vivo B, Lima A, Albanese S. Lead isotopes in soils and groundwaters as tracers of the impact of human activities on the surface environment: The Domizio-Flegreo Littoral (Italy) case study. J Geochem Explor 2011; 109: 51-8.
[http://dx.doi.org/10.1016/j.gexplo.2010.09.012]

[46]  Capra GF, Odierna PC, Coppola E, Grilli E. Occurrence and distribution of key potentially toxic elements (PTEs) in agricultural soils: a paradigmatic case study in an area affected by illegal landfills. J Geochem Explor 2014; 145
[http://dx.doi.org/10.1016/j.gexplo.2014.06.007]

[47]  Senior K, Mazza A. Italian "Triangle of death" linked to waste crisis. Lancet Oncol 2004; 5(9): 525-7.
[http://dx.doi.org/10.1016/S1470-2045(04)01561-X] [PMID: 15384216]

[48]  Pirastu R, Pasetto R, Zona A, *et al.* The health profile of populations living in contaminated sites: SENTIERI approach. J Environ Public Health 2013; 2013939267
[http://dx.doi.org/10.1155/2013/939267] [PMID: 23853611]

[49]  Rivezzi G, Piscitelli P, Scortichini G, *et al.* A general model of dioxin contamination in breast milk: results from a study on 94 women from the Caserta and Naples areas in Italy. Int J Environ Res Public Health 2013; 10(11): 5953-70.
[http://dx.doi.org/10.3390/ijerph10115953] [PMID: 24217180]

[50]  Pan-European regional assessment of the Sixth Global Environmental Outlook United Nations Environment Programme (Unep). Batumi, Georgia: United Nations Economic Commission For Europe (Unece) 2016.8-10

[51]  Penzo S. Il radon e la sua misurazione ENEA 2006.

[52]  13th International Congress of the International Radiation Protection Association, 13-18 May 2012, Glasgow, Scotland. http://www.irpa.net/images/stories/irpa12/irpa%2013%20bid%20(uk).pdf

[53]  Guida D, Guida M, Cuomo A, Guadagnuolo D, Siervo V. Assessment and Mapping of Radon-prone Areas on a regional scale asapplication of a Hierarchical Adaptive and Multi-scale Approach for the Environmental Planning. Case Study of Campania Region, Southern Italy. WSEAS Transactions on Systems 2013; 2: 105-20.

[54]   Neznal M, Hůlka J. Radon-Prone Areas – Some Remarks. Proceedings of the European Conference on Protection against Radon at Home and at Work. Praha. 1997; pp. 152-5.

[55]   Darby S, Hill D, Auvinen A, *et al.* Radon in homes and risk of lung cancer: collaborative analysis of individual data from 13 European case-control studies. BMJ 2005; 330(7485): 223.
[http://dx.doi.org/10.1136/bmj.38308.477650.63] [PMID: 15613366]

[56]   Nero AN, Nazaroff WW. Characterising the source of Radon indoors. Radiat Prot Dosimetry 1984; 7: 23-39.
[http://dx.doi.org/10.1093/oxfordjournals.rpd.a082958]

[57]   Gunderson LCS, Reimer GM, Agard SS. Correlation between geology, radon in soil gas, and indoor radon in the Reading Prong.Geological causes of natural radionuclides anomalies Special publication 4. Missouri, Dept. Natural Resources. Div. Geology & land survey 1988; pp. 91-102.

[58]   OwenDE. Relationships between Geology, equivalent Uranium concentration, and Radon in soilgas. Schuman RR. U.S. Geological Survey Open-File ReportFairfax County, Virginia 1988; 88: pp. 18-28.

[59]   Bolinder AS, Owen DE, Schumann RR. A preliminary evaluation of environmental factors influencing day-to-day and seasonal soil-gas radon concentrations. Field studies of radon in rocks, soil, and environment. USGS. C. K. Smoley Publishers 1993; pp. 23-31.

[60]   Choubey VM, Bist KS, Saini NK, Ramola RC. Relation between soil-gas radon variation and the different lithotectonic units of Garhwal Himalaya, India. J Appl Rad Isotop 1999; 51(5): 587-92.
[http://dx.doi.org/10.1016/S0969-8043(98)00149-3]

[61]   Choubey VM, Bartarya SK, Ramola RC. Radon in himalayan springs: A geohydrological control. Environ Geol 2000; 39(6): 523-30.
[http://dx.doi.org/10.1007/s002540050463]

[62]   Choubey VM, Bartarya SK, Saini NK, Ramola RC. Impact of geohydrology and neotectonic activity on radon concentration in groundwater of intermontane Doon Valley, Outer Himalaya, India. Environ Geol 2001; 40(3): 257-66.
[http://dx.doi.org/10.1007/s002540000177]

[63]   Davey G, Newport M. Podoconiosis: the most neglected tropical disease? Lancet 2007; 369(9565): 888-9.
[http://dx.doi.org/10.1016/S0140-6736(07)60425-5] [PMID: 17368134]

[64]   Skinner HCW. The earth, source of health and hazards: an introduction to medical geology. Annu Rev Earth Planet Sci 2007; 35: 177-213.
[http://dx.doi.org/10.1146/annurev.earth.34.031405.125005]

[65]   AGM Italia. Geologia Medica e salute umana in Italia, workshop 2011.
http://www.agmitalia.org/2011/Workshop_Geologia_Medica_2011.pdf

[66]   De Vivo B, Lima A, Siegel F. 2 Geochimica ambientale - Metalli potenzialmente tossici. Liguori Editore Napoli 2004; p. 446.

[67]   Bortone I, Erto A, Chianese S, Di Nardo D. Risk Analysis for a Contaminated Site in North of Naples(Italy) In: Pierucci Sauro, J Jiri, Klemes, Eds. Chemical Engineering Transaction, Chapter: Risk Analysis for a Contaminated Site in North of Naples (Italy). Publisher: AIDIC ServiziS.r.l 2015; pp. 1927-32.

[68]   Albanese S, Tafani MVE, De Vivo B, Lima A. An environmental epidemiological study based on the stream sediment geochemistryof the Salerno province (Campania region, Southern Italy). J Geochem Explor 2013; 131: 59-66.
[http://dx.doi.org/10.1016/j.gexplo.2013.04.002]

[69]   De Vivo B, Lima A, Albanese S, Cicchella D. Atlante geochimico-ambientale della Regione Campania/GeochemicalEnvironmental Atlas of Campania Region. Roma: Aracne Editrice 2006; p. 216.

[70]    Albanese S, De Luca ML, De Vivo B, Lima A, Grezzi G. Chapter 16. Relationships Between Heavy Metal Distribution and Cancer Mortality Rates in the Campania Region, Italy Environmental GeochemistrySite Characterization Data Analysis and Case Histories. 2008; pp. 387-400.
[http://dx.doi.org/10.1016/B978-0-444-53159-9.00016]

[71]    Ducci D, Albanese S, Boccia L, *et al.* An Integrated Approach for the Environmental Characterization of a Wide Potentially Contaminated Area in Southern Italy Int J Environ Res Public Health 2017 Jun; 2714(7): E693.
[http://dx.doi.org/10.3390/ijerph14070693]

[72]    Minolfi G, Albanese S, Lima A, Tarvainen T, Fortelli A, DeVivo B. A regional approach to the environmental risk assessment - Human health risk assessment case study in the Campania region. Journal of Geochemical Exploration184, Part B 2018 January; 400-16.

[73]    Rezza C, Albanese S, Ayuso R, Lima A, Sorvari J, De Vivo B. Geochemical and Pb isotopic characterization of soil, groundwater, human hair, and corn samples from the DomizioFlegreo and Agro Aversano area (Campania region, Italy). J Geochem Explor 2017.
[http://dx.doi.org/10.1016/j.gexplo.2017.01.007]

[74]    Comba P, Bianchi F, Fazzo L, *et al.* Health impact of waste management campania working group. Cancer mortality in an area of Campania (Italy) characterized by multiple toxic dumping sites. Ann N Y Acad Sci 2006; 1076: 449-61.
[http://dx.doi.org/10.1196/annals.1371.067] [PMID: 17119224]

[75]    Cicchella D, De Vivo B, Lima A, Albanese S, Mcgill RAR, Parrish RR. Heavy metal pollution and Pb isotopes in urban soils of Napoli, Italy Geochemistry: Exploration, Environment, Analysis. 2008; 8: pp. (1, 01.02)103-12.

[76]    Tarvainen T, Albanese S, Birke M, Poňavič M, Reimann C. TGP team arsenic in agricultural and grazing land soils of Europe. Appl Geochem 2013; 28: 2-10.

[77]    Civitillo D, Ayuso RA, Lima A, *et al.* Potentially harmful elements and lead isotopes distribution in a heavily anthropized suburban area: the Casoria case study (Italy). Environ Earth Sci 2016; 75(19): 1.
[http://dx.doi.org/10.1007/s12665-016-6093-4]

[78]    Rezza C. Multimedia environmental geochemical characterization Of DomizioFlegreo And Agro Aversano Area (Campania Region, Italy) Baselines, Isotopic Ratiosand Advanced Health Risk Assessment Phd Thesis 2017.

[79]    Albanese S, Fontaine B, Chen W, *et al.* Polycyclic aromatic hydrocarbons in the soils of a densely populated region and associated human health risks: the Campania Plain (Southern Italy) case study. Environ Geochem Health 2015; 37(1): 1-20.
[http://dx.doi.org/10.1007/s10653-014-9626-3] [PMID: 24908325]

[80]    De Felip E, Bianchi F, Bove C, *et al.* Priority persistent contaminants in people dwelling in critical areas of Campania Region, Italy (SEBIOREC biomonitoring study). Sci Total Environ 2014; 487: 420-35.
[http://dx.doi.org/10.1016/j.scitotenv.2014.04.016] [PMID: 24797738]

[81]    Giaccio L, Cicchella D, De Vivo B, Lombardi G, De Rosa M. Does heavy metals pollution affect semen quality in men? A case of study in the metropolitan area of Naples (Italy). J Geochem Explor 2012; 112: 218-25. https://www.sciencedirect.com/science/journal/03756742/112/supp/C

[82]    Bhattacharya S, Gupta K, Debnath S, Ghosh UC, Chattopadhyay D, Mukhopadhyay A. Arsenic bioaccumulation in rice and edible plants and subsequent transmission through food chain in Bengal basin: a review of the perspectives for environmental health. Toxicol Environ Chem 2012; 94: 429-41.
[http://dx.doi.org/10.1080/02772248.2012.657200]

[83]    Naujokas MF, Anderson B, Ahsan H, *et al.* The broad scope of health effects from chronic arsenic exposure: update on a worldwide public health problem. Environ Health Perspect 2013; 121(3): 295-302.

[http://dx.doi.org/10.1289/ehp.1205875] [PMID: 23458756]

[84]    Wardrop NA, Le Blond JS. Assessing correlations between geological hazards and health outcomes: Addressing complexity in medical geology. Environ Int 2015; 84: 90-3.
[http://dx.doi.org/10.1016/j.envint.2015.07.016] [PMID: 26233556]

# CHAPTER 7

# Environmental and Human Health Issues in Campania Region Italy

**Gabriella Marfe**[1,*], **Carla Di Stefano**[2] and **Professor Giulio Tarro**[3]

[1] *Department of Scienze e Tecnologie Ambientali, Biologiche e Farmaceutiche, University of Campania "Luigi Vanvitelli," via Vivaldi 43, Caserta t81100, Italy*

[2] *Department of Hematology, "Tor Vergata" University, Viale Oxford 81, 00133 Rome, Italy*

[3] *Primario emerito dell'Azienda Ospedaliera "D. Cotugno", Napoli, Italy*

**Abstract:** Waste generation rates continue to grow around the world, creating a need for more comprehensive waste management strategies to meet sustainability needs. Uncontrolled disposal generates complex and challenging situation that involves the entire population. In particular, the illegal dumping and burning of toxic waste in Campania (Italy) has caused immense environmental damage and an increase in cancer rate among the population. Different epidemiological studies were commissioned by the Ministry of Health to assess the magnitude of contamination under an illegal dump in Campania and to evaluate the population health impact. The data and other available evidence testify the dramatic situation in Caserta and Naples provinces about the severe impairment of the environmental conditions in several places and increase of cancer incidence. For this reason, the Campania region is known, such as Triangle of Death" and "Land of Fires" (LoF) (or Terra dei fuochi-TdF), as reported both in academic publications and the national press. This chapter is aimed to provide the findings regarding human health and environmental contamination in this region.

**Keywords:** Campania, Hazardous waste, Illegal waste dumping, Land of fire, Population health.

## RECENT HISTORY

At the opening of the judicial year on January 26, 2013, the late attorney general of Naples, Dr. Vittorio Martusciello had requested an investigation into the relationship between spills and tumor growth, focusing on the chronic environmental emergency that has been scourging Naples and province for years and on waste disposal and counterfeiting in the agro-food sector. The prosecutor

---

* **Corresponding aouthor Gabriella Marfe:** Department of Scienze e Tecnologie Ambientali Biologiche e Farmaceutiche, University of Campania "Luigi Vanvitelli," *via* Vivaldi 43, Caserta 81100, Italy; Tel: (+39) 0823 275104 ; Fax: +39 0823 274813 ; E-mail: gabriellamarfe@gmail.com

also stated that he did not feel reassured about the correlation between aggression to the environment and cancer diseases as they would have expected in Rome (statements by the Minister of Health Balduzzi) [1]. In July 2012, the book "Campania, land of poisons" was presented, and it dealt the illegal spill of toxic waste that had led to an increase in cancer and birth defects [2]. We do not always know that all the substances contained in illegally spilled toxic waste, but the presence of some outlines the inevitable carcinogenic effects. It is very difficult to find the correlation between the incidence of cancer and environmental cases because of the lack of a regional cancer registry for mapping cancer mortality cases in Campania. Genetic damage, through the malfunctioning of the switches of the genes, can be caused by exposure to carcinogens with mutagenic action (which cause changes in the germinal DNA). Epigenetics is now dealing with all this as a new interpretation of the increased rate of cancers. Already in 2004, K. Senior and A. Mazza had explored the possible effects of environmental pollution on cancer-related deaths in the Nola area, publishing the results in the prestigious journal "The Lancet Oncology". The territories of Campania produce a greater quantity of waste than that of landfills and incinerators and, therefore, their non-disposal leads, inevitably, to an increase in the incidence of cancer cases [3]. In an epidemiological study published in 2009 by the scientific Journal Clinical and Experimental Cancer Research, the data obtained from the national archive of hospital discharge records for the period between 2000 and 2005 were analyzed: the number of breast cancer was greater than 40,000 cases compared to that reported by the official bodies with underestimated statistics of 26.5% and were also affected age groups between 25 and 44 years [4]. In 2011, the American Journal of Cancer Biology and Therapy published the scientific results of a research that highlighted a significant increase in cancer deaths and congenital malformations in the Campania region where toxic waste had been disposed (including arsenic, mercury, dioxins and furans) in an unsafe manner [5]. In this regard, 30 years of non-properly disposed waste cost a mortality rate equal to 9.2% more for men in North Naples and South Caserta. 12.4% more for women [6]. After the publication of "Campania, land of poisons", another paper reported the increased number of quadrantectomies of mammary tumors in female patients between 25 and 39 years and between 40 and 44 years, *i.e.* in pre-screening age [7]. At the end of 2013, another article, published in Cancer Biology and Therapy found the increased mortality between 1988 and 2009 in the metropolitan areas of Naples and Caserta for different kinds of cancer [8]. Therefore, it is possible to respond positively to the questions on the incidence of tumors and mortality in Campanian territories, being greater than the Italian average. Naturally, scientific rigor and a map of polluted sites are needed. We know of numerous carcinogens to which the pollution caused by dioxins is added, but the greatest danger is the pollution of the aquifer layer linked to illicit spills (heavy metals). After several interventions on the correlation between environmental pollution and cancer and the certainly affirmative answer to the question on the existence of an increase in the incidence of tumors linked to the disposal of toxic waste, amply demonstrated by scientific publications, we must point out that the situation of Irpinia and the rest of Campania is not different from that present in the land of fires (Naples and Caserta). At this point, we believe it is important to elaborate ISTAT (Istituto Nazionale di Statistica) data on life expectancy, with a comparison between the Campania provinces and the Italian average. In this regard, in 1992, the Italian average of life expectancy (males) was 74 years, while in 2010, it became 79.4 years. In Campania, it has gone from 73.2 to 77.8 years, therefore from a minus 0.8 to a doubling of less 1.6 (Table 1). Furthermore, the risks associated with exposure to asbestos were evaluated at the Isochimica of Avellino to avoid asbestos contamination of other people. The data in Campania on increased mortality due to cancer are scary (when compared to other regions) because, in this region, there is a very serious environmental compromise [2].

Table 1. Life Expectation (ISTAT data).

| Years of Average Life in Campania | | | | |
|---|---|---|---|---|
| MALE | 1992 | DIFFERENCE | 2010 | DIFFERENCE |
| ITALY | 74.0 | 0 | 79.4 | 0 |
| CAMPANIA | 73.2 | -0.8 | 77.8 | -1.6 |
| AVELLINO | 75 | +1.0 | 79.2 | -0.2 |
| BENEVENTO | 74.8 | +0.8 | 79.0 | -0.4 |
| CASERTA | 72.2 | -1.8 | 77.4 | -2.0 |
| NAPLES | 72.3 | -1.7 | 77.2 | -2.2 |
| SALERNO | 74.6 | +0.6 | 78.5 | -0.9 |

The abusive spills of urban and industrial waste remain, especially in the suburbs, starting from the fast roads, in the entry ramps, a stone's throw from the virtuous municipalities of waste sorting, under each bridge, in every night. The acrid smoke that you breathe on the highway as you cross the boundary of the boundless suburb of Naples is more eloquent than any sign [2]. The uncontrolled fires and the burial of heavy metals, asbestos, cadmium, continue as if nothing had happened, in our region, in particular in the area of the Aversa where they continue to pollute aquifers layers and agricultural products [2]. The latest data available on the emissions register is managed by ISPRA (Istituto Superiore per la Protezione e la Ricerca Ambientale) and date back to a few years ago. They are the following and concern only emissions into the atmosphere: the industry contributes to PM 10 emissions for 26%, 70% of sulfur oxides, 23% nitrogen oxides, which are also secondary PM 10 precursors and ozone. As micro-pollutants, the atmospheric emissions of the industry are as follows: benzene 15%, IPA 34%, nickel 35%, total cadmium 60%, dioxins 70%, mercury 74%, lead 83%, PCB 86%, chromium 89% and arsenic 98%. In absolute terms, the PM 10 amounted to 180 thousand tons, 56 as chrome and 110 as mercury. The dioxins produced by the sector amount to 225 g. With regard to waste, the source is the annual ISPRA Report. In 1997, the first year of the Ronchi Decree, urban waste amounted to 26.6 million tons, in 2001, 29 million, and, with the last figure referring to 2011, it is 32.5 million tons (Table **2**). Special waste from 72 million tons in 2000 to 117 in 2006 to 118.2 (we have been in crisis for 4 years). At least 20 million special wastes disappear into thin air (Report of the Parliamentary Committee on the waste cycle): 85% of special waste is produced in 4 regions: Lombardy, Veneto, Emilia Romagna and Piemonte (Table **2**).

**Table 2.  Difference between urban wastes and special waste. WASTE (ISPRA Annual Report). 85% of special waste is produced in Lombardy, Veneto, Emilia Romagna, Piedmont.**

| Urban Waste | | |
|---|---|---|
| (1st year Ronchi decree) 1997 | 26.6 | Millions of tons |
| 2001 | 29 | Millions of tons |
| 2011 | 32.5 | Millions of tons |
| Special Waste | | |
| 2000 | 72 | Millions of tons |
| 2006 | 117 | Millions of tons |
| (after 4 years of "crisis") | 118.2 | Millions of tons |

## EPIDEMIOLOGICAL INVESTIGATIONS (U.S. NAVY)

As reported by the Naval Support Activity Naples [9], drinking water contamination was detected in the tap water of homes from unauthorized private wells and to a much lesser extent than those using a public drinking water source. Areas that seem to be influenced by the emission of chemical agents in the soil and/or aquifer layers (dioxins, furans, pesticides, polychlorinated biphenyls, metals, mercury vapor and aldehydes) have been identified, as a statistically significant linear trend was found in the increase of persistent asthmatics since 2006 in the staff of the USN (United States Navy Department), while these tendencies have not been found in Rota in Spain or in Sigonella in Sicily [2]. In response to the concern of US personnel about the potential health consequences of waste management practices in the Campania region, the navy commander for Europe, Africa and Southeast Asia region requested the execution of a public health assessment by the center for public health of the corps of the Navy. Samples of a series of media were detected, such as air, tap water, soil and soil gases on which analyzes were performed for 241 chemical agents and microorganisms (for instance, total and faecal coliforms) [2].

## THE LAND OF FIRES

In the census of sites potentially contaminated by dioxin (normal values 3 pg per gram of soil), large areas of the provinces of Naples and Caserta are polluted. In some areas of Campania, such as Acerra and Cercola, peaks of 50 and more pg of dioxin (in Seveso, for 49.6 pg the Army with specialized departments for reclamation) were measured in the ground.

**Fig. (1).** Land of fires-terra dei fuochi.

As for the sites contaminated as a result of illegal disposal, the practice of burning the areas of illegal waste disposal is very common. The provinces of Caserta and Naples have become the dustbin of Europe: the areas exploited for legal and illegal waste disposal are always the same. As reported in "Campania, land of poisons" there is the graphic overlap of the route of the state highway 162 (median axis) with respect to the municipal areas at greater risk of cancer and neonatal malformations identified by the so-called Bertolaso's study of 2007. The illegal spill of toxic sewage, the burning of waste (not for nothing today we talk about the land of fires), the infiltration into the subsoil are now sadly recurrent events in the media and in the daily experience of hundreds of thousands of citizens who they feel their future and their children's lives so seriously compromised. Still, many inhabitants of these territories feel the situation more tragically because they have to deal with a higher incidence of tumors within their own family (Fig. 1). The region Campania was called Campania Felix [Prosperous Campania] by the Romans because of the extreme fertility of north/north-east of the city of Naples area, renowned for its many local food products [10, 11]. Today, the demand of food products is drastically decreased on international markets because of the massive environmental devastation of the eco-system of the air, water resources and agricultural fields. Furthermore, the important role in agriculture of this area was challenged, as many grazing animals such as sheep and buffalo started to get ill. Initially, the public health authorities refused the relationship between the high cancer rates and the illegal toxic waste dumpsites, also due to the lack of reliable data at a medical level. In 2004, things started to change after the publication of article written by Senior and Mazza in 2004 as reported above. Another article reported the high death rate of the sheep in Acerra that was being affected by the toxic dumping [4]. As toxic elements, especially dioxin, sheep breeding in the area were the first victims of this dramatic situation. In this regard, the sheep were the first indicators of the disaster since their deaths were caused by the presence of enormous dioxin levels in the feed and in the whole countryside of Acerra. Furthermore, shepherds also started to die because of the high average of dioxin in blood, and this happened in a zone without any significant industrial activity. In this scenario, several studies were performed in order to find a link between toxic waste and the high incidence of different types of cancer. In a review, Triassi *et al.* [12] observed a possible relationship between the long-term exposure of waste and liver and lung cancer in the population living in the Caserta and Naples provinces. Altavista *et al.* [13] reported that the mortality rate among citizens living in Giugliano, Villaricca and Qualiano municipalities (51 km$^2$) was higher when compared with that expected on the basis of the regional data as provided by ISTAT. A high mortality ratio was found for a different type of cancer among the male and female populations of Giugliano (+7.23% and +11.08%, respectively), but not in the population of Villaricca and Qualiano. Importantly, in this specific area (108 km$^2$), according to data provided by ARPA (Agenzia Regionale per la Protezione Ambientale Campania http://www.arpacampania.it/), Campania and Commissioner for Environmental Emergency and Lega Ambiente Campania, 39 waste disposal sites were identified: 2 authorized and 37 illegal sites. The data reported the presence of twenty-seven of the 39 waste that were contaminated with toxic wastes. One study [14] conducted on the population living in the 196 municipalities of the provinces of Naples and Caserta (approximately 4 million citizens) reported a statistically significant increase of cancer mortality rate in the provinces of Naples (+8.7% in men and +9.2% in women) and Caserta (+2.3% in males). In this context, the Dipartimento della Protezione Civile (Department of Civil Protection) commissioned a study on the treatment of waste and the health status of populations resident in the neighbourhood of toxic wastes dumping sites in Campania. The study, performed by different institutions (WHO, CNR, Italian Institute of Health (ISS), Regional Epidemiological Observatory (Osservatorio Epidemiologico Regionale) ARPAC (Regional Agency for Environmental

Protection of Campania-Azienda Regionale per la Protezione Ambientale della Campania) and l'ESA (Epidemiology Development and the Environment-Epidemiologia per lo Sviluppo e l'Ambiente) showed that there was a higher excess of mortality among females (23% in Province of Caserta and 47% in Province of Naples) than males (19% in Province of Caserta and 47% in Province of Naples) and an increased rate of different types of cancer such as kidney, liver, trachea, lungs, bladder, and pleura [15]. Another research calculated the environmental contamination by a computing waste exposure at a municipal level using the database of legal waste landfills and illegal dumping sites after adjustment for social and economic factors in the 196 municipalities of the provinces of Caserta and Naples [16]. Such sites were characterized by the regional Environmental Protection Agency and were divided into seven categories on the basis to their potential contamination respect to water, soil and air. Each municipality was considered to be affected by one dumping site located within a 1 km radius and each of them was numbered from 1 to 5 by increasing risk (1 = least at risk; 5 = most at risk). Furthermore, this study measured a municipal deprivation index on the basis of five variables: the primary education in the population, the average surface of dwellings, the unemployment rate among the population, the percent of the population not owning their dwelling and the percent of mono-parental families, after adjustment for social and economic factors analysis. Significant excess relative risk estimates were found in the highest group for both cancer mortality (4.1% and 6.6% in males and females, respectively) and for liver cancer mortality (19.3 and 29.1% in males and females, respectively) when compared with the lowest group [16]. The other two studies identified different clusters of cancer in several municipalities [17, 18]. Specifically, the first study [17] found two clusters of lung cancer in the central part of Naples province and three clusters of liver cancer (two of them in the northern part of the Province of Naples and one in the southern part of the Province of Caserta). The second study [18] reported a high incidence for several kinds of cancer such as liver, lung, non-Hodgkin's lymphoma, leukaemia in both sexes; testis, esophagus, larynx, pancreas, thyroid gland, and morphologic soft-tissue sarcomas in men; kidney, biliary ducts, brain cancer and myeloma in women, and finally, stomach and topographic soft tissue sarcomas in the overall population. In line with these studies, another project, called Sentieri, showed high cancer mortality in 44 highly polluted national sites. In particular, the authors found an excess of mortality for liver cancer in both sexes and laryngeal cancer in men, in the 77 municipalities of the Naples and Caserta provinces (Litorale domizio flegreo e agro aversano) [19, 20]. Specifically, the authors reported an excess of hospitalized children with cancer (leukemia and Central Nervous System-CNS Tumors) in the first year of life in the Caserta provinces. In 2019, the update of SENTIERI Project detected an excess for many diseases in people living close to National Priority Contaminated Sites (NPCSs). The methodological approach used this study did not allow to adjust for several confounding factors since they represented risk factors for the considered diseases (*e.g.*, smoking, alcohol consumption, obesity). Furthermore, the environmental characterization of the studied NPCSs was not uniform in terms of quality and detection of the pollutants, because of the lack of information in different databases. It was calculated an excess of cancer incidence of 1,220 case in men and 1,425 in women over a five-year time window. Specifically, an increased incidence of malignant mesothelioma, lung, colon, and gastric cancer, and for non-malignant respiratory diseases was observed. in people living close to NPCSs with the presence of chemical and petrochemical plants, oil refineries, and dumping hazardous wastes. Furthermore, the authors also observed an excess of non-malignant respiratory disease in people living close to NPCSs with steel industries and thermoelectric plant, while an increase of mesothelioma was observed in NPCSs characterized by the presence of asbestos and fluoro-edenite, For the first time, the study reported the health status of children and adolescents (1,160,000 subjects, aged 0-19 years), and young adults (660,000 subjects, aged 20-29 years). The authors found high hospitalizations (about 7,000) among infants (0-1 year) (2,000 for perinatal origin). Furthermore, an elevated rate of hospitalizations (22.000 in the age class 0-14) was observed for different causes (4,000 due to acute respiratory diseases, and 2,000 to asthma). The authors found an excess of cancer incidence in the age group 0-29 years (9%) for different cancers such as soft tissue sarcomas in children (aged 0-14 years), acute myeloid leukaemia in children (aged 0-29 years), non-Hodgkin lymphoma and testicular cancer in young adults (aged 20-29 years). Additionally, an increased rate of overall congenital anomalies (genital organs, heart, limbs, nervous system, digestive system, and urinary system at birth) was observed in seven out of 15 considered NPCSs. Finally, the authors pointed out the importance of communication among the population living close to NPCSs, considering their cultural frame and their network of relationships [21]. In another recent article, the authors reported the number of cancer cases registered by the AIEOP (Association Italian Pediatric Hematology-Oncology) in 25 years (1990–2014). During this time, 3655 cancers were observed (Napoli province-2059 cases, Salerno province-625, Caserta province-589 Avellino province-229 and Benevento province-153) [22], The authors underlined that the overall ratio between observed (O) and expected (E) numbers of cases in these five periods increased from 0.69 in the first period to 0.76, then 0.82, 0.91, and 0.94, in the other considered periods [22]. In the last decades, strong and unexplained increases in the incidence of childhood tumors were observed in Europe, especially in Italy, as reported by a review published in Lancet Oncology [23]. The last official data on pediatric tumors in Campania (2012) were in accordance with this study (Table **3**), (provided by the

Local Sanitary District of Naples 1 (ASL1) Napoli http://www.tumori.net/ banche_dati/tumori/query.php) (Piano Regionale di Programmazione Della Rete Ospedaliera ai sensi del DM 70/2015 2016-2018).

Table 3. Cancer incidence and mortality trend in children and adults Campaniaa *vs* Italy.

| Childhood Cancers | | | |
|---|---|---|---|
| *Cases* per *100000 inhabitants* | *Cases* per *100000 inhabitants* | *Differences between Italy and Campania* | % increment rate of cancer |
| 0-14 years of age | 0-14 years of age | | |
| 1993-Campania | 2012-Campania | | |
| 10.8 | 17,3 | 6,5 | 60 |
| 1993-Italy | 2012-Italy | | |
| 13,9 | 18,9 | 5 | 35 |
| Cancer incidence rate in adults | | | |
| *Cases* per *100000 inhabitants* | | | |
| 2010 | | | |
| Men -Campania | Men-Italy | | |
| 395 | 336 | 62 | 15,5 |
| Cancer Mortality Rates in adults | | | |
| *Cases* per *100000 inhabitants* | | | |
| 2012-Campania | 2012-Italy | | |
| Men | Men | | |
| 231 | 173 | 58 | 25 |
| Women | Women | | |
| 103 | 94 | 9 | 8,7 |

Furthermore, the EPIKIT study [24] has found in Campania (2007-2011) about 3.465 cases of cancers in pediatric age (0-19 years). Another study reported that 439 patients with lung cancer underwent surgery at the Thoracic Surgery Division of the National Institute of Cancer in Naples between November 2004 and April 2013. 123 patients came from the land of the fires (28%), 316 (72%) came from other areas (Benevento, Avellino, Salerno). Following surgery for lung cancer, the patients that lived in TdF had long term survival very similar to patients that lived in other areas. For this reason, the authors concluded that their study could not establish a causal link between the presence of toxic waste and' onset of lung cancer [25]. Moreover, since a lot of hazardous waste was illegally burned by releasing dangerous chemicals (such as 2,3,7,8-tetraclhorodibenzo-*p*-dioxin-

-TCDD), other studies were carried out on this population to measure these compounds by using different approaches. One study measured PCDDs, PCDFs and DL/NDL-PCBs level in the blood serum of 25 subjects from Acerra, Nola, Marigliano and 33 subjects from the city of Naples and its surroundings (Pompei, Portici, Pozzuoli, Torre del Greco) [26]. The authors found that higher and lower substances levels were respectively associated with age (> 50-year-old woman) and obesity. Another study [27] analyzed milk samples in a sample of 95 primiparas from 41 municipalities of the provinces of Naples and Caserta (who had given birth to newborns at the Caserta Hospital between June 2007 and May 2008). Forty-eight samples were derived from women living in municipalities with a low waste index score, while 47 samples from women living in municipalities with a high waste index score. Environmental dioxin risk score was correlated with the waste index score (Spearman rank correlation coefficient $\varrho$ = 0.275; $p < 0.0001$). In another paper, the authors reported that PCDD-PCDF levels were significantly correlated to the presence of illegal burning of hazardous waste in the living area of the sampled women examined [28]. Another article [29] calculated the dioxins concentrations in the breast milk of mothers living in Giugliano (Campania, Italy), (where there were many illegal burning of solid waste) and in two cities in northern Italy, Milan and Piacenza, which were chosen as controls. The total PCDD/F and DL-PCB TEQs in samples derived from Giugliano were significantly lower than those derived from Milan and Piacenza. In this case, the authors observed a significant correlation between the mother's age and an increased concentration of PCDDs/Fs and PCBs in their milk. However, a correlation between some type of cancer and individuals overexposed to 2,3,7,8-TCDD or to dioxins, in general, is still controversial and yet to be demonstrated [30 - 33]. In a larger biomonitoring SEBIOREC (Studio Epidemiologico Biomonitoraggio Regione Campania), the authors analyzed the highly toxic persistent contaminants in blood, blood serum, and human milk of a large number of healthy donors in Naples and Caserta provinces. Many substances (such as Polychlorodibenzodioxins (PCDDs), polychlorodibenzofurans (PCDFs), and polychlorobiphenyls (PCBs, dioxin-like (DL) and non-dioxin-like ($\Sigma$6PCBs)), arsenic (As), cadmium (Cd), mercury (Hg), and lead (Pb) were measured in serum (organic biomarkers) and blood. The authors observed that the PCDDs, PCDFs, and DL-PCBs were sample-discriminative, whereas NDL-PCBs were the least effective, possibly for the limited number of congeners assayed. Furthermore. the municipalities of Caserta province were most marked by the organic contaminants rather than to Naples province. On the contrary, metals levels were high in Brusciano, Caivano, Giugliano, Mugnano, and Qualiano–Villaricca of Naples province, above all the presence of As. Furthermore, Hg concentration was elevated in different municipalities of Naples. Such a province is a typical volcanic territory, where not uncommon geological formations contain As and

Hg. Moreover, the authors also reported unexpected results for the municipalities of both provinces: for example, Castel Volturno and Caivano had very similar levels of PCDD, PCDF and DL-PCB congeners in comparison with the other municipalities of the same province and near to one another. Therefore, the analysis of biomarker distribution can potentially play a key role in future local risk assessment and/or management actions [34]. Other assessments of dioxins was performed in milk samples from sheep, buffalo, and cattle raised in the Campania region and it showed that about one fourth of them presented levels of dioxins much higher than the nor-mal threshold indicated by European Commission (3.0 pg TEQ/gfat) [35-38]. Another study found an increased pre-valence of lymphoma in dogs living close to illegal waste dumps [39]. Basile *et al.* [40] conducted a study on mosses collected in the town of Acerra, and they reported different ultrastructural abnormalities at the level of cytoplasm vesicles and concentric multilamellar/multivesicular bodies. They hypothesized that could possibly occur as a response to an adaptive mechanism because of environmental heavy metal pollution. Other authors were carried out research [41] on a specific type of frogs that are a good bioindicator for detecting genotoxic effects of chemical, environmental hazards. Specifically, they showed severe DNA damage in the erythrocytes of frogs living in north Campania. In accordance with these data, another article reported shorter telomere length and lower telomerase activity in peripheral blood mononuclear cells derived from women that were subjected to therapeutic abortion in the second trimester of pregnancy and living close illegal waste sites [42]. Bergamo *et al.* [43] carried out a pilot study (EcoFoodFertility initiative) on human semen in 110 healthy males living in various areas of Campania with either high or low environmental impact. The authors found higher zinc, copper, chromium and reduced iron levels, as well as reduced sperm motility and higher sperm DNA Fragmentation Index (DFI) in the semen of men that living in a very polluted area. Another pilot biomonitoring study was conducted in 112 clinically healthy, normospermic men living in various areas of the Campania region (South of Italy) with high or low environmental pressure on telomere length (TL) in both leukocytes (LTL) and sperm cells (STL).

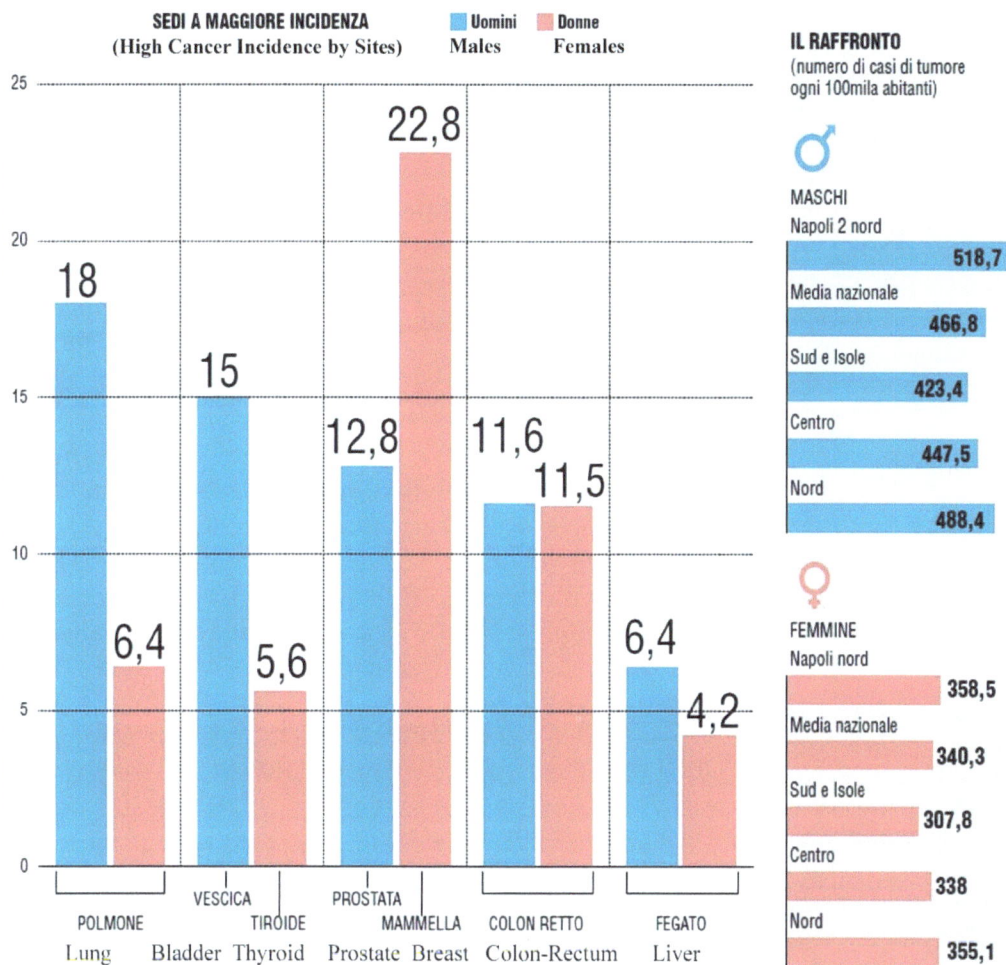

**Fig. (2).** Cancer incidence provided from ASL 2 North Naples, The Cancer Registry of the A.S.L. North includes 32 municipalities in the Naples province for a total of 1,051,000 inhabitants including 32 municipalities in the Naples province for a total of 1,051,000 inhabitants. The Cancer Registry of the A.S.L. North includes 32 municipalities in the Naples province for a total of 1,051,000 inhabitants including 32 municipalities in the Naples province for a total of 1,051,000 inhabitants. Cancers number was 518 per 100 thousand inhabitants in males and 358.1 per 100 thousand in females in North Naples 2 (ASL 2).

In this study, the authors found that telomeres were longer in sperm cells derived from young people living in areas of high environmental pressure when compared with those derived from young people living in areas of low environmental pressure [44]. Another review [45] reported the high incidence of bladder cancer in the Naples province with respect to the rest of Europe. In another recent project, named Epica (Project EPIdemiologia Cancro) (2015), the authors have found an increased incidence of different types of cancer when compared to national average during 2012-2013 (breast + 2%, prostate (+ 3%), bladder +1%, tyroid +1%, lung +2%, liver +2% and brain +2%) in people living in Casoria (municipality of Naples). In particular, the authors observed that population growth decreased by 2.5% during 2008-2014 respect to 2003-2007 with an increase of new cancer diagnosis during the same time [46]. The similar results have been reported by the study Perseo, conducted in Terzigno: in this municipality of Naples, it was observed an increase of different types of cancer (lung ca, leukemia and in particular, an increasing number of children and young adults with central nervous system tumors [47]. Furthermore, both studies will proceed with population-based cancer screening in a primary care network in order to design a map of incidence and mortality. Another study detected the high incidence of different types of cancer in the Naples and Caserta provinces with respect to the rest of Europe [48]. Moreover, the data provided from Local Sanitary District of North Naples 2 (ASL 2), reported an elevated rate of different kinds of cancer such as liver, breast, lung, bladder, colon-rectum, prostate and thyroid in both sexes when compared with regional and national average. Such sanitary District of North Naples includes 32 municipalities of the Naples province for a total of 1,051,000 inhabitants. In this area, 13,792 new cases of cancer were registered from 2010 to 2012 (7,632 cases (55%) among males and 6,160 cases (45%) among females. The most frequent cancers were lung (18%), bladder (15%) prostate (12.8%), colorectal (11.6%) and liver (6.4%) in men, while in women were: breast (28.2%), colorectal (11.5%), lung (6.4%), thyroid (5.6%) and liver (4.2%) (Fig. **2**). Furthermore, 6760 deaths due to cancer were registered (4086 males and 2674 females) up a total of 21,066 deaths for all causes from 2010 to 2012. These results showed an increased cancer incidence that was statistically significant when compared to data of the central, the south Italy (including other areas of Campania, Sardina and Sicily) and to the national average. However, this value was not statistically significant when compared to that of the north Italy (Fig. **2**) [49]. In another article, Petrosino et al. reported the presence of metals and PCBs in a small sample of cancer patients in two different matrices such as blood and hair. Quantitative analyses were performed at samples of blood and hair for 14 heavy metals and at blood samples for 12 PCBs. The author found that the half of the 14 heavy metals concentration was higher in the blood of 11 cancer patients when compared to the reference values, while the levels measured in hair presented higher values than the maximum reference.

Moreover, some patients showed higher positive values of the 12 PCBs in blood samples when compared to the maximum tabular reference (although there is no clear reference quantified in the WHO-2005 report). Furthermore, the authors observed a light increase about 3 out of 14 heavy metals in in blood samples of patients with head and neck cancer. Moreover, these patients had high PCB concentrations in their blood samples respect to the maximum reference level.

The heavy metals concentration in hair were also double in comparison to the maximum reference [50]. Another very recent study of the same team was conducted on the blood and hair of 33 cancer patients (living for at least 10 years in areas at high risk of pollution of Campania and Basilicata), to evaluate the presence of heavy metals and PCBs. The authors calculated the concentration of 14 metals (aluminum, antimony, arsenic, barium, cadmium, chromium, iron, lithium, mercury, nickel, lead, copper, strontium and zinc) and 12 PCBs. They reported that the levels of most metals were increased in all cancer patients. Interestingly, a male patient with the testicular seminoma and female patients with ductal carcinoma of the breast had an increased level of these metals. Nevertheless, no association were found between the levels of metals and the stage of cancer. Elevated levels of certain PCBs were also found in most patients. Although the precise role of these substances in carcinogenesis is still unclear, further studies are necessary specifically to compare cancer patients in areas with high or low environmental impact to better understand the effects of these substances on cancer risk at the population level [51]. Although numerous epidemiological studies have shown the high incidence of cancer in Campania, new study designs and procedures have to incorporate biological/biochemical tools.

## CONCLUSION

In Campania, the epidemiological surveillance in contaminated areas could play a crucial role to monitor changes in health status related to exposure to environmental contamination through modern methodologies and credible information. There are different factors that have to be considered in this dramatic situation (i) the scarcity of resources including finance, equipment, personnel and data for waste planning; (ii) lack of enforcement of existing regulations on toxic solid waste and urban environmental management in general.

In this scenario, it should be established new rules on environmental protection following different steps: (a) to introduce higher levels of differential waste; (b) to improve the waste selector efficiency, (c) to eliminate the illegal burning and waste sites through major control of the legal forces in the territory; (d) to begin the screening programs on exposed human populations for toxicological chemical detection, ;and (e) to activate research centers in Campania to better characterize the pollution types and sources and to elaborate measures to reduce associated health risks. Furthermore, this Region has started new environmental monitoring campaigns on the food chain and soil and water contamination. At the end, the national government delegates and regional authorities under the Europe for Citizens Program (COHEIRS: Civic Observers for Health and Environment: Initiative of Responsibility and Sustainability- ruled by ALDA (Association for

Local Democracy European agency in Strasbourg, the International Society Doctors for the Environment (ISDE), and ISBEM (Istituto Scientifico Biomedico Euro Meditteraneo research center) have involved the population in monitoring and preventing the illegal activities of burning or dumping, ruled. There is still room in Italy to adopt many changes in order to prevent future environmental injustices. Furthermore, since the enforcement of waste control regulations remains week, it is desirable that legislators and Institutions would make many efforts to increase public participation in environmental decision-making

## CONSENT FOR PUBLICATION

Not applicable.

## CONFLICT OF INTEREST

The authors confirm that this chapter content has no conflict of interest.

## ACKNOWLEDGEMENTS

Declare none.

## LIST OF ABBREVIATIONS

| | |
|---|---|
| **NDL/DL-PCBs** | non dioxin/dioxin-like PCBs |
| **PCBs** | polychlorinated biphenyls |
| **PCDDs** | Polychlorinated dibenzodioxins |
| **PCDFs** | polychlorinated dibenzofurans |
| **TEQ** | toxicity equivalence |

## REFERENCES

[1]   Anno giudiziario l'inaugurazione. Il Pg: rifiuti e tumori, le rassicurazioni non ci convingono. Corriere del Mezzogiorno, 27 gennaio 2013. Judicial year the inauguration. Pg: waste and tumors, assurances do not convince us. Corriere del Mezzogiorno, 27 January. http://www.giuliotarro.it/ wordpress/ incidenza-aumentata-dei-tumori-nella-terra-dei-veleni

[2]   Giordano A, Tarro G. Campania, Terra di veleni. Denaro Libri, Naples - July 2012 https://www.lafeltrinelli.it/libri/antonio-giordano/ campania-terra veleni/ 9788874440672?productId= 9788874440672.html

[3]   Senior K, Mazza A. Italian "Triangle of death" linked to waste crisis. Lancet Oncol 2004; 5(9): 525-7. [http://dx.doi.org/10.1016/S1470-2045(04)01561-X] [PMID: 15384216]

[4]   Piscitelli P, Santoriello A, Buonaguro FM, *et al.* CROM Human Health Foundation Study Group Incidence of breast cancer in Italy mastectomies and quadrantectomies performed between 2000 and 2005. J Exp Clin Cancer Res 2009; 19: 28- 86. [http://dx.doi.org/10.1186/1756-9966-28-86] [PMID: 19545369]

[5]     Barba M, Mazza A, Guerriero C, *et al.* Wasting lives: the effects of toxic waste exposure on health. The case of Campania, Southern Italy. Cancer Biol Ther 2011; 12(2): 106-11. Epub 2011 Jul 15. [http://dx.doi.org/10.4161/cbt.12.2.16910] [PMID: 21734464]

[6]     Greyl L, Vegni S, Natalicchio M, Cure S. Ferretti J Waste Crisis in Campania, Italy A SUD http://www.ceecec.net/case-studies/waste-crisis-in-campania-italy/

[7]     Piscitelli P, Barba M, Crespi M, *et al.* . Human Health Foundation Study Group, in memory of Prof. Giovan Giacomo The Burden of Breast Cancer in Italy: Mastectomies and Quadrantectomies Performed Between 2001 and 2008 based on Nationwide Hospital Discharge Records. J Exp Clin Cancer Res 2012; 20: 31-96.

[8]     Crispo A, Barba M, Malvezzi M, *et al.* Cancer mortality trends between 1988 and 2009 in the metropolitan area of Naples and Caserta, Southern Italy: Results from a joinpoint regression analysis. Cancer Biol Ther 2013; 14(12): 1113-22. [http://dx.doi.org/10.4161/cbt.26425] [PMID: 24025410]

[9]     Naval support activity naples about health awareness. https://www.cnic .navy.mil/ regions/cnreurafcent/installations/nsa_naples/about/health_awareness.html2008.

[10]    Iacuelli A. Le vie infinite dei rifiuti. Il sistema campano Rinascita Rome 2008. https://www.libroco.it/dl/Alessandro-Iacuelli/Rinascita/9788890325427/Le-v-e-infinite-dei-rifiuti-Il-sistema-campano/cw627981789070634.html

[11]    Iovene B. Campania Infelix. Editore: HYPERLINK https://www.ibs.it/ libri/editori/bur-bibliotec--univ.-rizzoli BUR Biblioteca Univ. Rizzoli Collana: HYPERLINK https://www.ibs.it/ libri/collane/futuropassato-p207812 Futuropassato Anno edizione: 2008 https://www.ibs.it/campania-infelix-libro-bernardo-iovene/e/9788817026352

[12]    Triassi M, Alfano R, Illario M, Nardone A, Caporale O, Montuori P. Environmental pollution from illegal waste disposal and health effects: a review on the "triangle of death". Int J Environ Res Public Health 2015; 12(2): 1216-36. [http://dx.doi.org/10.3390/ijerph120201216] [PMID: 25622140]

[13]    Altavista P, Belli S, Bianchi F, *et al.* Cause of specific mortality in a district of Campania Region with high number of waste dump sites. Epidemiol Prev 2004; 28: 311-21. [PMID: 15792153]

[14]    Comba P, Bianchi F, Fazzo L, *et al.* Health Impact of Waste Management Campania Working Group. Cancer mortality in an area of Campania (Italy) characterized by multiple toxic dumping sites. Ann N Y Acad Sci 2006; 1076: 449-61. [http://dx.doi.org/10.1196/annals.1371.067] [PMID: 17119224]

[15]    WHO Regional Office for Europe *et al.* "Trattamento dei rifiuti in Campania: impatto sulla salute umana. Studio Pilota. Malformazioni congenite nelle province di Napoli e Caserta (1996-2002): analisi descrittiva e struttura spaziale del rischio", 2005; Id., "Trattamento dei rifiuti in Campania: impatto sulla salute umana. Studio Pilota. Mortalità per tumori nelle province di Napoli e Caserta (1994-2001): analisi descrittiva e struttura spaziale del rischio", 2005 (both documents are no longer available on the Civil Protection's website, but a synthesis of both is found on the ISS epidemiological observatory's website, EpiCentro: http://www.epicentro.iss.it/temi/ambiente/notepernapoli26012005.pdf, "Tratta mento dei rifiuti in Campania: impatto sulla salute umana. Studio Pilota. Sintesi dei risultati e indicazioni preliminari", accessed 20 October 2015)

[16]    Martuzzi M, Mitis F, Bianchi F, Minichilli F, Comba P, Fazzo L. Cancer mortality and congenital anomalies in a region of Italy with intense environmental pressure due to waste. Occup Environ Med 2009; 66(11): 725-32. [http://dx.doi.org/10.1136/oem.2008.044115] [PMID: 19416805]

[17]    Fazzo L, Belli S, Minichilli F, *et al.* Cluster analysis of mortality and malformations in the Provinces of Naples and Caserta (Campania Region). Ann Ist Super Sanita 2008; 44(1): 99-111. [PMID: 18469382]

[18]   Fazzo L, De Santis M, Mitis F, *et al.* Ecological studies of cancer incidence in an area interested by dumping waste sites in Campania (Italy). Ann Ist Super Sanita 2011; 47(2): 181-91.
[PMID: 21709388]

[19]   Pirastu R, Zona A, Ancona C, *et al.* [Mortality results in SENTIERI Project]. Epidemiol Prev 2011; 35(5-6) (Suppl. 4): 29-152. https://www.ncbi.nlm.nih.gov/pubmed/24986501
[PMID: 22166295]

[20]   Pirastu R, Ricci P, Comba P, *et al.* Mortality results in SENTIERI Project. Epidemiol Prev 2011; 35: 29-15. [Italian].
[PMID: 24986501]

[21]   Zona A, Iavarone I, Buzzoni C, *et al.* Gruppo di lavoro SENTIERI; Gruppo di lavoro AIRTUM-SENTIERI; Gruppo di lavoro Malformazioni congenite-SENTIERI. SENTIERI: Epidemiological Study of Residents in National Priority Contaminated Sites. Fifth Report]. Epidemiol Prev 2019; 43(2-3 Suppl 1): 1-208. Italian
[http://dx.doi.org/doi: 10.19191/EP19.2-3 S1.032] [PMID: 31295974]

[22]   Indolfi P, Picazio S, Perrotta S, *et al.* IICC-3 contributors. Time trends of cancer incidence in childhood in Campania region: 25 years of observation. Ital J Pediatr 42016: 2-82.
[http://dx.doi.org/10.1186/s13052-016-0287-y]

[23]   Steliarova-Foucher E, Colombet M, Ries LAG, *et al.* International incidence of childhood cancer, 2001-10: a population-based registry study. Lancet Oncol 2017; 18(6): 719-31.
[http://dx.doi.org/10.1016/S1470-2045(17)30186-9] [PMID: 28410997]

[24]   Piscitelli P, Marino I, Falco A, *et al.* Reply to the Letter of Terracini B. *et al.* "Comment on Piscitelli *et al.* Hospitalizations in Pediatric and Adult Patients for All Cancer Type in Italy: The EPIKIT Study under the E.U. COHEIRS Project on Environment and Health". Int. J. Environ. Res. Public Health 2017, 14, 495. Int J Environ Res Public Health 2017; 14(11): E1291.
[http://dx.doi.org/10.3390/ijerph14111291] [PMID: 29068388]

[25]   Rocco G, Petitti T, Martucci N, *et al.* Survival after surgical treatment of lung cancer arising inthe population exposed to illegal dumping of toxic waste in the land of fires ('Terra dei Fuochi') of Southern Italy. Anticancer Res 2016; 36: 2119-24.

[26]   Esposito M, Serpe FP, Diletti G, *et al.* Serum levels of polychlorinated dibenzo-p-dioxins, polychlorinated dibenzofurans and polychlorinated biphenyls in a population living in the Naples area, southern Italy. Chemosphere 2014; 94: 62-9.
[http://dx.doi.org/10.1016/j.chemosphere.2013.09.013] [PMID: 24112656]

[27]   Rivezzi G, Piscitelli P, Scortichini G, *et al.* A general model of dioxin contamination in breast milk: results from a study on 94 women from the Caserta and Naples areas in Italy. Int J Environ Res Public Health 2013; 10(11): 5953-70.
[http://dx.doi.org/10.3390/ijerph10115953] [PMID: 24217180]

[28]   Giovannini A, Rivezzi G, Carideo P, *et al.* Dioxins levels in breast milk of women living in Caserta and Naples: assessment of environmental risk factors. Chemosphere 2014; 94: 76-84.
[http://dx.doi.org/10.1016/j.chemosphere.2013.09.017] [PMID: 24120012]

[29]   Ulaszewska MM, Zuccato E, Capri E, *et al.* The effect of waste combustion on the occurrence of polychlorinated dibenzo-p-dioxins (PCDDs), polychlorinated dibenzofurans (PCDFs) and polychlorinated biphenyls (PCBs) in breast milk in Italy. Chemosphere 2011; 82(1): 1-8.
[http://dx.doi.org/10.1016/j.chemosphere.2010.10.044] [PMID: 21074246]

[30]   Murray IA, Patterson AD, Perdew GH. Aryl hydrocarbon receptor ligands in cancer: friend and foe. Nat Rev Cancer 2014; 14(12): 801-14.
[http://dx.doi.org/10.1038/nrc3846] [PMID: 25568920]

[31]   Perucatti A, Di Meo GP, Albarella S, *et al.* Increased frequencies of both chromosome abnormalities and SCEs in two sheep flocks exposed to high dioxin levels during pasturage. Mutagenesis 2006;

21(1): 67-75.
[http://dx.doi.org/10.1093/mutage/gei076] [PMID: 16434450]

[32]   Bodner KM, Collins JJ, Bloemen LJ, Carson ML. Cancer risk for chemical workers exposed to 2,3,7,8-tetrachlorodibenzo-p-dioxin. Occup Environ Med 2003; 60(9): 672-5.
[http://dx.doi.org/10.1136/oem.60.9.672] [PMID: 12937189]

[33]   Bertazzi PA, Consonni D, Bachetti S, *et al.* Health effects of dioxin exposure: a 20-year mortality study. Am J Epidemiol 2001; 153(11): 1031-44.
[http://dx.doi.org/10.1093/aje/153.11.1031] [PMID: 11390319]

[34]   De Felip E, Bianchi F, Bove C, *et al.* Priority persistent contaminants in people dwelling in critical areas of Campania Region, Italy (SEBIOREC biomonitoring study). Sci Total Environ 2014; 487: 420-35.
[http://dx.doi.org/10.1016/j.scitotenv.2014.04.016] [PMID: 24797738]

[35]   Santelli F, Boscaino F, Cautela D, Castaldo D, Malorni A. Determination of polychlorinated dibenzo-p-dioxins (PCDDs), polychlorinated dibenzo-p-furans (PCDFs) and polychlorinated biphenyls (PCBs) in buffalo milk and mozzarella cheese. Eur Food Res Technol 2006; 223: 51-6.
[http://dx.doi.org/10.1007/s00217-005-0112-0]

[36]   Pecoraro A. Activity of Veterinary Services of ASL-NA 4 during the Dioxins Emergency Regione Campania, Azienda Sanitaria ASL-NA4, Servizio Veterinario, Igiene Degli Alimenti e Produzione Zootecniche. Naples, Italy: ASL-NA4 2005. https://www.epicentro.iss.it/discussioni/rifiuti/pecoraro

[37]   Mazza A, Piscitelli P, Neglia C, Della Rosa G, Iannuzzi L. Illegal Dumping of Toxic Waste and Its Effect on Human Health in Campania, Italy. Int J Environ Res Public Health 2015; 12(6): 31-6818.

[38]   Marfe G, Di Stefano C. The evidence of toxic wastes dumping in Campania, Italy. Crit Rev Oncol Hematol 2016; 105: 84-91.
[http://dx.doi.org/10.1016/j.critrevonc.2016.05.007] [PMID: 27424919]

[39]   Marconato L, Leo C, Girelli R, *et al.* Association between waste management and cancer in companion animals. J Vet Intern Med 2009; 23(3): 564-9.
[http://dx.doi.org/10.1111/j.1939-1676.2009.0278.x] [PMID: 19298612]

[40]   Basile A, Sorbo S, Aprile G, *et al.* Heavy metal deposition in the Italian "triangle of death" determined with the moss *Scorpiurum circinatum*. Environ Pollut 2009; 157(8-9): 2255-60.
[http://dx.doi.org/10.1016/j.envpol.2009.04.001] [PMID: 19446383]

[41]   Maselli V, Polese G, Rippa D, Ligrone R, Kumar Rastogi R, Fulgione D. Frogs, sentinels of DNA damage induced by pollution in Naples and the neighbouring Provinces. Ecotoxicol Environ Saf 2010; 73(7): 1525-9.
[http://dx.doi.org/10.1016/j.ecoenv.2010.05.011] [PMID: 20684845]

[42]   De Felice B, Nappi C, Zizolfi B, *et al.* Telomere shortening in women resident close to waste landfill sites. Gene 2012; 500(1): 101-6.
[http://dx.doi.org/10.1016/j.gene.2012.03.040] [PMID: 22465532]

[43]   Bergamo P, Volpe MG, Lorenzetti S, *et al.* Human semen as an early, sensitive biomarker of highly polluted living environment in healthy men: A biomonitoring study on trace elements in blood and semen and their relationship with sperm quality and RedOx status. Reprod Toxicol 2016; 66: 1-9.
[http://dx.doi.org/10.1016/j.reprotox.2016.07.018] [PMID: 27592743]

[44]   Vecoli C, Montano L, Borghini A, *et al.* Effects of Highly Polluted Environment on Sperm Telomere Length: A Pilot Study. Int J Mol Sci 2017; 18(8): E1703.
[http://dx.doi.org/10.3390/ijms18081703] [PMID: 28777293]

[45]   Di Lorenzo G, Federico P, De Placido S, Buonerba C. Increased risk of bladder cancer in critical areas at high pressure of pollution of the Campania region in Italy: A systematic review. Crit Rev Oncol Hematol 2015; 96(3): 534-41.
[http://dx.doi.org/10.1016/j.critrevonc.2015.07.004] [PMID: 26520458]

[46]    PROGETTO EPI. Studio epidemiologico sul cancro nella città di Casoria Città di Casoria Distretto 43 asl Na 2 nord Medici e Pediatri di famiglia di Casoria Comune Casoria Distretto 43 Asl Napoli 2; February 2015. http://docslide.it/health-medicine/progetto-epica-report-riassuntivo-semplificato.html

[47]    Annunziata A, Esposito MR. Report Sanitario-Ambientale. 2015. https://it.geosnews.com/p/ it/campania/na/terzigno-pronti-progetti-per-il-risanamento-del-teritorio-e-dei-cittadini-- uesto-posto-non-pu-non-deve-morire_8407373

[48]    Mazza A, Piscitelli P, Falco A, *et al.* Heavy Environmental Pressure in Campania and Other Italian Regions: A Short Review of Available Evidence. Int J Environ Res Public Health 2018; 15(1): E105. [http://dx.doi.org/10.3390/ijerph15010105] [PMID: 29320415]

[49]    Mautone Ettore. Terra dei Fuochi piu Tumori edizione II Mattino 1 Giugno 2017. http://www.ilmattino.it

[50]    Petrosino V, Motta G, Tenore G, Coletta M, Guariglia A, Testa D. The role of heavy metals and polychlorinated biphenyls (PCBs) in the oncogenesis of head and neck tumors and thyroid diseases: a pilot study. Biometals 2018; 31(2): 285-95. [http://dx.doi.org/10.1007/s10534-018-0091-9] [PMID: 29520558]

[51]    Petrosino V, Coletta M, Testa D. Determination of Heavy Metals and Polychlorinated Biphenyls in Oncological Patients: A Pilot Study. Cancer Sci Res Open Access 2019; 6(1): 1-20.

# CHAPTER 8

# A Case Study on Grassroots Environmentalism for Health and Sustainability in the Land of Fires (Italy)

Salvatore Paolo De Rosa[*], Lucio Righetti and Annamaria Martuscelli

*Environmental Humanities Laboratory, Division of History of Science, Technology and Environment KTH Royal Institute of Technology, Stockholm, Sweden*

**Abstract:** This chapter provides a reassessment of the waste 'crises' in the so-called Land of Fires in Italy and highlights the contribution of local social mobilizations in advancing environmental justice and sustainability concerns. It is based on the long-term collaboration between a social scientist and two activists, and it draws from a dialogue of academic and activist knowledge in the tradition of Environmental Justice organizations and scholarship. Our contention is that local communities and ecologies of this area have borne the brunt of the socio-environmental costs of waste disposal and industrial production. This translated into risks of contamination for the residents, suffering threats to health and stigmatization. To this state of affairs, local grassroots environmental movements answered by confederating into a regional coalition that arranged on November 16, 2013, one of the biggest environmental justice demonstrations in Italian history, spurring the intervention of the State *via* legislation and resources. The coalition was also instrumental in the elaboration of the Pact for the Land of Fires, a programmatic commitment to end illegal waste disposal that defined the obligations for the institutional and civil society signatories. Finally, we show how the experience of activists is systematized and coordinated today through the initiative of the Civic Observers, an association of citizens that monitors environmental crimes and conditions in cooperation with public officials.

**Keywords:** Cancer, Civic observers, Environmental justice, Grassroots environmentalism, Hazardous waste, Organized crime, Waste management.

## INTRODUCTION

This chapter chronicles the key turning points of grassroots environmental organizing during the conflicts around waste management in the so-called Land of Fires of Campania region in Southern Italy, a phenomenon that has not received

---
[*] **Corresponding author Salvatore Paolo De Rosa:** Environmental Humanities Laboratory, Division of History of Science, Technology and Environment KTH Royal Institute of Technology, Stockholm, Sweden; E-mail: osservatoricivicicampania@gmail.com

enough attention so far. Moving from a reassessment of the waste 'crises' that traversed this region during the last two decades, our aim is to debunk reductionist readings of local communities' protests and proposals – frequent in mainstream renderings of environmental activism in Italy and elsewhere – by showing their actual contribution to health security, environmental governance and more sustainable human-environment relations. The chapter is the result of a long-term collaboration between one social scientist and two experienced activists, and it draws from a dialogue of academic and activist knowledge in the tradition of Environmental Justice organizations and scholarship [1 - 3].

The recent history of Campania cannot be grasped without due consideration of the longstanding social conflicts that emerged around waste (mis)management. Snapshots from Naples, the capital city, flooded by its municipal waste and reports of hazardous waste illegally disposed of proliferated in the news from about 2003 onwards [4]. International commentators, as well as Italian analysts, were bewildered. Mainstream media boiled down such a complex issue as the result of two main factors. Firstly, the inability of local communities to perform sorting and recycling of garbage associated with a reiteration of the NIMBY syndrome [5]. Secondly, the role played by organized crime groups hijacking the government's efforts towards a resolution of the management stall while driving people's protests [6, 7]. Particularly in the international media, the reference to "pre-existing ideas about Naples as an aberrant city on the margins of a 'normal' Europe" for deciphering the waste crisis, as Dines [4] argues in relation to British press, prevented an understanding of the complexities of the political, economic and administrative backgrounds while relying on, and reinforcing, dismissive stereotypes about southern Italian people.

These explanations hinted at alleged essential characteristics that would exclude southern Italians from civilized people. Such 'naturalization' of discontent was instrumental to the depoliticization of government's plans, paving the way for the violent repression of people's protests [8]. Additionally, by focusing on the *urban* side of the waste crisis – framed as a public order issue tainting Italy's reputation – political élites failed to promptly address the troubling reality of illegal disposal and burning of *hazardous* waste. Here, we offer a different diagnosis of the waste issue and provide insights into the role of the local grassroots mobilizations in advancing environmental justice and sustainability concerns. These insights are based on a doctoral dissertation on grassroots environmentalism in Campania by De Rosa [9] and on the personal and political experiences of Righetti and Martuscelli as key organizers of social mobilizations and grassroots advocacy. In our diagnosis, grassroots environmentalism in Campania constituted barriers to socioecological exploitation and elaborated more effective and sustainable paths. In the next section, we set the stage through an account of the waste 'crises' based on the merging of scientific scrutiny and the point of view of those who self-organized on the ground. Then, in the third section, we delve into the formation of the regional coalition *Stop Biocide*. In the fourth section, we present an initiative aimed towards environmental monitoring and public health put forward by the *Civic Observers for Environment and Health* and their role within the institutional framework of the *Pact for the Land of Fires*. In the fifth section, we describe the rise and the tasks of the *Civic Observers of Campania*, before advancing our conclusions.

## THE WASTED LAND

Campania's conflicts around waste played out mostly in the plain between Naples and Caserta, within the ancient *Campania Felix*, as the Romans called it. This area gained the appellation of *Terra dei Fuochi,* Land of Fires, given by activists to denounce the frequent uncontrolled fires of hazardous waste next to fertile cultivated fields [10]. Two processes revolving around the commodification of waste materials, and impacting negatively local socio-environmental conditions, dominated the recent history of this area. Firstly, the setting up of the Regional Urban Waste Management Plan through a special government agency operating under the framework of the 'waste emergency'. Secondly, the illegal trafficking and dumping of hazardous waste in improper landfills and agricultural areas organized by a criminal alliance of industrial managers, public officials, landowners and mafia groups [11, 12]. The complex interrelations between these processes almost make clear boundaries between legal and illegal practices of waste management disappear. This is best exemplified by the observable reduction of spaces of democratic confrontation on environmental governance [13]. In practice, these processes have shifted onto local communities and ecologies, the socio-environmental costs of waste disposal and manufacturing industrial activities, producing in turn risks of contamination affecting the health of local populations and causing the reduction, and stigmatization, of cultivated lands [14].

In 1994, the Italian government declared a 'waste emergency' in Campania. Contextually, it established the normative framework of the 'emergency regime'. From that point on, a Commissioner with extraordinary powers was invested with the task of supervising the implementation of the plan. The official motivation of the emergency was to increase controls and hasten operations. However, turning all private landfills into public ones for dissociating waste management from organized crime proved useless when the implementation and management of the waste cycle were assigned to a private company able to perform subcontracting without accountability. By 2009, when the emergency was officially ended, not only were the initial objectives not achieved, but also an increase in unequal exposure to environmental pollution and hazards had materialized.

Fifteen years of emergency established a sort of 'permanent' state of exception, legitimizing authoritarian rule, repression of social discontent and scarce environmental protection. The regulatory interventions enabled by the state of exception ultimately supported the economic interests of the corporation responsible for arranging the waste cycle. The Commissioner delivered in 1997 the regional plan for the management of about 2,500,000 tons of regional garbage produced annually [15]. The plan was based on the construction of two waste-t--energy facilities, seven plants for producing Refuse-Derived Fuel (RDF) from municipal garbage, and an unspecified number of landfills. Through a *project financing* scheme, the project had to be assigned to a private service provider through a public bid. In 2000, a group of private companies led by the corporation Impregilo won the contract. Subsequently, Impregilo exercised pressure on the Commissioner for achieving exceptions to the contract and privileges for its operations with the aim of increasing the potential economic gains from waste management and incineration [16]. Indeed, the corporation was allowed to store processed urban garbage ahead of the construction of the incinerator and to locate facilities in Campania through private land deals without consulting with local communities. The logistical needs of the corporation were privileged by designating the landscape as a simple backdrop for its profit-oriented operations. It is by considering the undemocratic core of this approach that one starts to understand better the rise of widespread resistance against the top-down imposition of waste facilities.

In Italy, incineration of waste is subsidized within a national scheme of renewable energies' incentives. Producing electricity by burning waste allows tapping in these public subsidies. With the aim of maximizing its profits, Impregilo transformed into RDF all the trash it could. Crucially, the RDF needed to be stored before the completion of the incinerator. Ultimately, sixteen temporary storage sites were built by combining privately-owned plots of land of various sizes. These sites accommodated between 6 and 7 million tons of *ecoballe,* the popular name for RDF. In 2019, after more than ten years from the setting up, most of these facilities are still storing the same garbage. According to calculations by Armiero and D'Alisa [17, 18], if the accumulated RDF was burned, it could have generated around one billion euros from subsidies. The latest Parliamentary Commission on Waste [19] reported that the costs for the public to lease the land where the waste is stored, as of 31 December 2017, amount to 23,793,587 euros.

The other, hidden side of the 'waste crisis' has its roots in the late 1980s, when Campania became the final destination of huge quantities of hazardous waste through illegal networks, allowing some industries and manufactures (mostly from Northern Italy) to shift the operation costs of their businesses on the environments and bodies of Campania residents. In 2000, the economic turnover of this illegal business was worth around 7.5 billion euros per year [20]. Currently, it is estimated to be almost 17 billion euros per year [21]. About 10 million tons in 22 years of hazardous waste arrived illegally in the region, according to Legambiente [21], ending up in toxic fires, in underground deposits, in urban waste landfills, on agricultural fields and in the concrete of apartment buildings and roads. If, on the one hand, the networks of this illicit system had national and international ramifications, on the other, the practice of illegal disposal and burning of discards and scraps from manufacturers operating from *within* Campania has proliferated.

The trafficking of hazardous waste towards Campania peaked in the 1990s and until the beginning of the 2000s. The *Camorra* clans were able to command the authorizations for incoming special waste through corrupt regional councilors [12]. Disposal sites were ensured by formally legal entrepreneurs, managers of landfills and landowners, while the network of customers included industrial groups and companies looking for cheap disposal options. The clans provided territorial hegemony and the local monopoly violence. As a result, millions of tons of hazardous waste were managed illegally. These included ashes from fumes abatement of steel plants, sludge from chemical processes, hospital waste and scraps of industrial productions, and many more hazardous materials.

The impacts on the local socio-environmental conditions caused by hazardous and urban waste are still disputed. In particular, the disagreements between the government, the experts and the grassroots movements regard the exact distribution of pollution and its links with local health conditions [22]. When the Regional Plan for Remediation of Polluted Sites of Campania Region was published in 2005, it recorded 2507 potentially contaminated sites [23], brought to 3733 in an update in 2009 [24]. The flocks of grazing animals raised in the region have been a 'sentinel' of environmental contamination: in 2003, about 12,000 cattle, river buffalos and sheep were culled for the levels of dioxin in their milk [25]. The reportage by Senior and Mazza [26], on the connection between mortality rates and the presence of dumping sites, popularized the label 'triangle of death' for the area between Acerra, Nola and Marigliano. But it was in 2004 that more solid evidence of the detrimental health effects of waste dumping in Campania became available, with the publication of a study focused on cancer incidence and mortality within 196 municipalities in the provinces of Naples and Caserta, commissioned by the Department of Civil Protection [27]. The results confirmed statistically significant correlations between health and toxic waste, with particularly worrying levels of specific diseases and malformations near sites of hazardous waste dumping. Moreover, several dedicated studies by the Italian Higher Institute of Health found an excess, in relation to the national average, of neonatal malformations and mortality for leukemia, sarcoma and malignant tumors of the lung, pleura, larynx, bladder, testicle, liver and brainstem [28 - 30]. Human chromosomal alterations caused by environmental damage were also documented [31] together with breast milk contamination from dioxin, the concentration of which directly depending on age, place of residence and exposure to toxic fires [32, 33]. Finally, the S.E.N.T.I.E.R.I study of 2014, focusing on 55 towns in the provinces of Naples and Caserta, found higher mortality and cancer incidence in the population, highlighting the worrying situation for children between 0 and 14 years [34]. A recent review of scientific research on the health effects of waste exposure and human biomonitoring in Campania by Triassi *et al.* [35] reiterates the connection between waste and increasing cancer rates in the region.

## Stop Biocide and the 'Raging River'

Civic involvement in Campania had the merit of highlighting the illegal disposal of hazardous waste and the consequent risks for public health. Social mobilizations have imposed the problem to public attention and have worked towards its resolution. This story may seem marginal if it was not, as we hope to demonstrate, emblematic of the role of grassroots activism in the context of late capitalism in a developed country, currently in the grip of the double climate and ecological crisis.

During June 2013, three of the main environmental movements of Campania built an alliance. The regional coalition *Stop Biocide* grew out of the agreement between the groups of 'Coordination of Committees against Toxic Fires' (or CCF), the 'Commons Network', and the 'Campania Citizens for an alternative waste management plan'. The CCF was the expression of secular and catholic local committees that emerged in the provinces of Naples and Caserta. Their central concern addressed the plague of illegal disposal of industrial and urban waste in the countryside often put on fire. Commons Network is an organized political collective with the primary objective of defending and supporting the commons, that was strengthened by the struggle of the local community of Chiaiano, in the north of Naples, against the opening of a landfill imposed by the central government in 2008. The Campania Citizens is a free association of citizens, committees and movements active in the city of Naples. It began its activities by opposing the government-sponsored waste management plan centered on landfills and incinerators with alternative solutions, based on the pillars of 4 Rs: Reduce, Reuse, Recycle, Redesign. Thanks to the support of public figures active in promoting alternative waste management schemes, like

Paul Connet and Rossano Ercolini, they managed to spread the principles of circular economy applied to waste.

That same year, the threat of construction of a new incinerator in the municipality of Giugliano, essentially aimed to dispose of the waste blocks of RDF (or *ecoballe*) stored in 'temporary' storage sites, triggered the reaction of the local population. Shortly after, the local committees, that emerged in the region during the years of struggle, joined the fight. The pyramids of waste blocks were the perfect embodiment of the failure of the regional plan: unsorted garbage, mixed with hazardous materials initially destined to the incinerator of Acerra, which materialized with great delay for judicial and technical problems. Still today, those blocks are mostly laying in the same places, as a memory of the failure of a waste cycle geared at processing, storing, landfilling and burning garbage for private profit.

The scars of those wounds were still aching for which reason few people could stand idly by the threat of a new incinerator. The Coalition started a campaign of capillary spreading of information, to make local communities aware and to explain the reasons for the protest, also showing that another way was possible. Sixty-four marches and demonstrations were organized in many municipalities of the Land of Fires, and many more meetings were held in clubs, churches, schools, private homes, squares and streets to warn local communities of the threats to health. The perception of these threats linked to the destruction of the environment conceived as a relational whole of air, land and water, became more and more clear, mobilizing hundreds of thousands of people in few months, from June to November 2013.

Every rally and demonstration in the Land of Fires was characterized by the high variety of the social composition: from the doctors for the environment (ISDE) to priests, workers, students, families, old and young people, as well as the deceased, represented by big photographs on banners as a way of remembering them. The pain of the mothers who lost their children contributed to raising the indignation of large parts of the local society. It appeared strange, to say the least, the widespread rising of cancer diseases among children that certainly could not be said of having 'unhealthy' lifestyles. With time, activists had to familiarize themselves with terms like incidence, mortality, prevalence and risk, that is, with the indicators of the statistical data on specific types of tumors. There was a widespread social perception that in the area of the Land of Fires, partially covered by the cancer register of ASL NA 3 SUD, the rates of incidence and mortality were higher compared to other, more industrialized and polluted areas of Italy [36].

This data alarmed the local communities, triggering protests for the lack of answers by public institutions for requesting clarity and transparency about the real risks to health and the countermeasures put into being. Nothing happened from the side of local, regional and national governments. Therefore, it was determined by the Coalition the need to "invade" Naples in order to show, with a massive peaceful demonstration, the will of the people. The preparation of the event was characterized by precision, organization and diffusion. The date of November 16, 2013, became the objective toward which were directed the efforts of over 300 local committees and associations of the Land of Fires. Many events of preparation were launched: flash-mobs, assemblies, meetings, conferences, trying to include every group and instance towards November 16. *Raging River* became the label of this demonstration, coined during an assembly at the former *Asilo Filangieri* of Naples. The contribution of mass media and artists, musicians, actors, writers, opinion leader, was massive. Social media were invaded by short clips, banners, photos, drawings; the creativity exploded into a stream of communication online and offline, with banners, flags, leaflets. National radio and television started to be interested in the phenomenon. National and international journalists came to investigate. Several Campanian radios participated by transmitting activists' calls to participate. No one could say that they did not know. Finally, several plenary assemblies preceded the November 16 demonstration, during which the final common platform was written down, shared by all the movements, associations and groups that joined the coalition.

November 16, 2013, was a rainy day. Fearing defections, the organizers were anxiously waiting for the development of the participation by those who shared the platform. The starting point was next to Naples' central station, in Mancini square. The demonstration began. It was crowded since the beginning, despite the rain, and people marched peacefully and colorfully towards Plebiscito Square, the biggest and most iconic square in Naples. Awaiting them, there was a massive stage from which the common platform was read aloud. Many groups were able to tell about their struggles and their reasons for joining the demonstration, including groups from other parts of Italy involved in socio-environmental conflicts. Rage, indignation and pain were expressed on the stage, against the absent state in all its declinations. While the head of the demonstration was arriving in Plebiscito Square, the tail was moving from the starting point. In total, about one hundred thousand people took part, about two km and six hundred meters of people marching. This has been the biggest environmental justice demonstration in the history of Italy.

## THE COHEIRS

During the preparation of the *Raging River* demonstration, public institutions were alarmed by the massive participation of people to the preparatory rallies, knowing too well they were the ultimate target of protests. Ministries, parliamentarians and regional representatives launched requests of dialogues towards the mobilizing people, to understand better the requests coming from the bottom, or to check the strength of the movement. Some representatives and organizers of grassroots committees were contacted and presented with the possibility of having a fair confrontation to find shared solutions so that the popular instances would finally be listened to.

In those days, the NGO ALDA (Association of Local Democracy Agencies) promoted the project COHEIRS: *Civic Observers for Health and Environment: Initiative for Responsibility and Sustainability,* aiming to establish a network of civic observers for safeguarding the health and the environment and to verify the application of the 'precautionary principle' sanctioned by the Maastricht Treaty and by art.191 of the Treaty of European Union Functioning.

This project aimed to confederate grassroots environmental associations and to stimulate the participation by the citizenry to decision-making processes in relation to environmental issues. It was a new, emerging opportunity: the possibility of constructive dialogue between institutions and citizens, a tool of participatory democracy, innovative but at the same time extremely fragile. It was not easy to give credit, among activists, to the new course that it seemed was sponsored by governing institutions. Two of the authors of this chapter, Righetti and Martuscielli, had an instrumental role in coordinating the dialogue around the construction of such partnership, without giving up other tools of denunciation, like the audition at the Environmental Commission of the Italian Senate and the European Parliament in Brussels. A delegation of representatives from local committees participated in the dialogue around this project moving from the shared vision outlined in the Coalition platform. The aim was to orient the debate and center the focus on the right to live in a healthy and safe environment. This was the first systematic attempt to bring the governing institution on a field drawn by activists.

The first results arrived. The government emanated the Law Decree 136/2013 (subsequently turned into Law 6/2014), focused on the analysis and monitoring of pollution on agricultural fields of Campania. Activists deemed it insufficient, but it was the first step. The regional parliament of Campania delivered the regional law 20/2013 that intended to answer to the requests of municipal administrators, lacking for the most part effective measures to reduce and prevent illegal disposal and burning of waste. Finally, the Ministry of Interior appointed the vice-prefect Donato Cafagna with the role of coordinating actions against the hazardous fires in Campania. Thanks to his determination and the pressure of local communities, protocols of coordination among the over 90 municipalities comprised by Law within the Land of Fires were delivered, through the so-called 'Pact for the Land of Fires'.

This pact was signed on July 11, 2013. It represented a programmatic commitment in which the obligations of the following signatories were defined: the Ministry of Interior, the Governor of Campania, the presidents of the provinces of Naples and Caserta, the prefects of Naples and Caserta, the mayors of all the municipalities included in the pact, representatives from the Regional Environmental Protection Agency and the Local Health Agencies, and finally the representatives from grassroots committees and large environmental NGOs. One of the main aims was to promote the participation of citizens in the decision-making process to achieve transparent management of local territories and environments.

As the head of the operational unit of the Pact for the Land of Fires, the vice-prefect Donato Cafagna coordinated the actions to fill the administrative vacuum in contrasting the fires of hazardous waste. The task he attended was to identify the available legislative and administrative tools that could connect different institutions and agencies usually not well synchronized, delineating for each specific area of responsibility. However, his powers were merely for coordination, therefore he could not sanction the misbehavior of the signatories of the pact but simply suggest, according to the legislative framework of his task, actions and changes.

The Pact was formulated by organically connecting all existing protocols, and by giving content with concrete actions to the wide set of environmental laws of the Law 152/2006. Among the tools of the Pact, particularly important were the synergy among municipalities for territorial control and environmental monitoring, the checks to manufacturers, the realization of municipal waste collection sites, and finally the delivery of the website *Prometeo* to allow the sharing of information with citizens and their participation through reports on environmental problems.

## CIVIC OBSERVERS OF CAMPANIA

From the experience of COHEIRS, emerged the activity of the *Campania Civic Observers,* an initiative of associated citizens, some from environmental movements, willing to contribute to environmental protection by devoting their time and energy.

Inspired by the COHEIRS protocol that followed the precautionary principle, the Civic Observers focused their activity on the prevention of environmental pollution. Thanks to the cooperation with Donato Cafagna, and the concrete help of the SMA, a semi-public company owned by Campania Region, a mobile phone application was delivered to signal a georeferenced, real-time monitoring of the illegal disposal sites scattered in the Land of Fires, the sites of burning trash and the forest fires. Once a person alerts the institutions through the app, it is validated by the operators at the SMA and reaches the relevant government agency in charge of dealing with the alarm. The photographs from each alarm are sent to the municipalities territorially responsible and to firefighters and municipal police. Moreover, each reporting is stored in the database of SMA, contributing to drawing an updated map of landfills and fires in ninety municipalities of the Land of Fires.

The use of the app was a success. Thanks to the easy access, it became very common for any ordinary person to report on environmental issues with their mobile phone. A major task of Civic Observers was to monitor the actions performed by the local administrations to prevent and control illegal waste disposal and waste fires. The Pact for the Land of Fires had set the minimum requirements of the municipal administrators to contrast the toxic fires and to access the economic resources made available by the government. It was not easy to check on the over ninety municipalities that had signed the Pact. The municipalities were compelled by law to: establish a partnership with Ecopneus, the consortium of tire manufacturers; to implement periodic control on productive activities in order to register them and classify them according to their methods to dispose of scraps and discards; to control the construction sector, especially permits and waste disposal; to map and monitor the temporary waste storage sites; to establish municipal centers for collecting waste, the so-called ecological islands; and to monitor and remove all asbestos artifacts. Based on the difficulties in gathering such amount and variety of information, the Civic Observers came up with the idea of developing Audit procedures: periodic meetings between the pact signatories, the institutions and representatives from associations and movements that would go through all the responsibilities according to the law and their level of the actual implementation.

Finally, with support of the secretary and the scientific committee of the Civic Observers, and the imprimatur of the head of the Pact for the Land of Fires, young volunteers were trained on the normative aspects of the Pact, on the Regional Law 20/2013, and the more recent Regional Law 548/2016, to help them contrasting the phenomenon of illegal waste fires, besides checking on the correct application of protocols and laws of waste management by the municipal administrations. This set of actions, and the continuous work of Civic Observers, are contributing to establishing in Campania the necessary conditions of health security for local communities, pushing also for a transition towards more sustainable use of resources.

## CONCLUSION

This chapter has provided an overview of the waste 'crises' in Campania and a chronicle of recent turning points of grassroots environmental mobilizations, from the development of a Coalition of movements to the promotion of several bottom-up projects geared at the monitoring of environmental conditions and prevention of environmental crimes. Firstly, we presented a short account of the waste 'crises' from the point of view of critical researchers and activists. Secondly, we described the creation of the Coalition of movements and the organization of the biggest demonstration for environmental justice ever happened in Italy, held in Naples on November 13, 2013. Thirdly, we analyzed the rise and effects of the Pact for the Land of Fires, the COHEIRS and the Civic Observers, providing an account of the strategies put in motion and of other connected projects that grew out of the cooperation between activists, volunteers and institutions. Such wealth of bottom-up pro-active behavior by inhabitants and communities in the Land of Fires debunks reductionist and stereotypical accounts of social mobilizations in Campania. Far from being the 'cause' of waste mismanagement and stall of the system, the organized exercise of citizens' rights by the local communities provided fundamental instigation and support to the institutional activities concerning waste management and prevention of health risks. Their contribution in terms of sustainability has been invaluable, showing how the systemic involvement of citizens in all phases of environmental governance can contribute in fundamental ways in designing and carrying out sustainable management of resources, cities and society in general.

## CONSENT FOR PUBLICATION

Not applicable.

## CONFLICT OF INTEREST

The author declares that there is no conflict of interest in this chapter.

## AKNOWLEDGEMENTS

Salvatore Paolo De Rosa would like to thank FORMAS (Swedish Research Council for Sustainable Development) for providing funding for this research under the National Research Programme on Climate (Contract: 2017-01962_3).

## REFERENCES

[1]   Martinez-Alier J, Anguelovski I, Bond P, Del Bene D, Demaria F. Between activism and science: grassroots concepts for sustainability coined by Environmental Justice Organizations. J Polit Ecol

2014; 21(1): 19-60.
[http://dx.doi.org/10.2458/v21i1.21124]

[2]     Martínez-Alier J, Healy H, Temper L, *et al.* Between science and activism: learning and teaching ecological economics with environmental justice organisations. Local Environ 2011; 16(1): 17-36.
[http://dx.doi.org/10.1080/13549839.2010.544297]

[3]     Conde M. From activism to science and from science to activism in environmental-health justice conflicts. J Sci Commun 2015; 14(2): C04.
[http://dx.doi.org/10.22323/2.14020304]

[4]     Dines N. Bad news from an aberrant city: a critical analysis of the British press's portrayal of organized crime and the refuse crisis in Naples. Mod Italy 2013; 18(4): 409-22.
[http://dx.doi.org/10.1080/13532944.2013.801677]

[5]     http://www.economist.com/printedition/2008-01-122019.

[6]     Yardley   J.   https://www.nytimes.com/2014/01/30/world/europe/beneath-southern-italy-adeadly-mob-legacy.html2019.

[7]     Phillips                                                                                                    J. http://www.independent.co.uk/news/world/europe/prodi-under-pressure-asarmy-moves-in-to-clear-nap les-trash-768798.html2019.

[8]     Petrillo A. Le urla e il silenzio. Depoliticizzazione dei conflitti e parresia nella Campania tardo liberale.Biopolitica di un rifiuto Le rivolte anti-discarica a Napoli e in Campania. Verona: Ombre Corte 2009.

[9]     De Rosa SP. Reclaiming territory from below: Grassroots environmentalism and waste conflicts in Campania, Italy. Lund: Lund University 2017.

[10]    Caggiano M. De Rosa, S. P. Social Economy as Antidote to Criminal Economy: How social cooperation is reclaiming commons in the context of Campania's environmental conflicts. Partecipazione e Conflitto 2015; 8(2): 530-54.

[11]    Massari M, Monzini P. Dirty Businesses in Italy: A Case-study of Illegal Trafficking in Hazardous Waste, Global Crime, 2004; 6:3-4: 285-304.

[12]    Iacuelli A. Le vie infinite dei rifiuti 2007.

[13]    D'Alisa G, Walter M, Burgalassi D, Healy H. Conflict in Campania: waste emergency or crisis of democracy. Ecol Econ 2010; 70: 239-49.
[http://dx.doi.org/10.1016/j.ecolecon.2010.06.021]

[14]    De Rosa SP. A political geography of 'waste wars' in Campania (Italy): Competing territorialisations and socio-environmental conflicts. Polit Geogr 2018; 67: 46-55.
[http://dx.doi.org/10.1016/j.polgeo.2018.09.009]

[15]    D'Alisa G, Di Nola MF. Giampietro MA multi-scale analysis of urban waste metabolism: Density of waste disposed in Campania. J Clean Prod 2012; 35: 59-70.
[http://dx.doi.org/10.1016/j.jclepro.2012.05.017]

[16]    Rabitti P. Ecoballe. Roma: Aliberti 2008.

[17]    Armiero M, D'Alisa G. Rights of resistance: The garbage struggles for environmental justice in Campania, Italy. Capitalism Nat Socialism 2012; 23(4): 52-68.
[http://dx.doi.org/10.1080/10455752.2012.724200]

[18]    Armiero M, D'Alisa G. Voices, Clues, numbers: Roaming among waste in Campania. Capitalism Nat Socialism 2013; 24(4): 7-16.
[http://dx.doi.org/10.1080/10455752.2013.851262]

[19]    Parliamentary Commission on Waste. Relazione territoriale sulla regione Campania. XVII Legislatura 2018.

[20]  Parliamentary Commission on Waste. Seduta di mercoledí 25 ottobre 2000. Presidente Massimo Scalia. XIII Legislatura 2000.

[21]  Legambiente Terra dei fuochi: radiografia di un ecocidio. Roma: Legambiente 2013.

[22]  Cantoni R. The waste crisis in Campania, South Italy: a historical perspective on an epidemiological controversy. Endeavour 2016; 40(2): 102-13.
[http://dx.doi.org/10.1016/j.endeavour.2016.03.003] [PMID: 27180606]

[23]  ARPAC. Piano regionale di bonficia dei siti inquinati della regione Campania. Napoli 2005.

[24]  ARPAC. Relazione sullo stato dell'ambiente in Campania 2009. Napoli. 2009.

[25]  Perucatti A, Di Meo GP, Albarella S, *et al.* Increased frequencies of both chromosome abnormalities and SCEs in two sheep flocks exposed to high dioxin levels during pasturage. Mutagenesis 2006; 21(1): 67-75.
[http://dx.doi.org/10.1093/mutage/gei076] [PMID: 16434450]

[26]  Senior K, Mazza A. Italian "Triangle of death" linked to waste crisis. Lancet Oncol 2004; 5(9): 525-7.
[http://dx.doi.org/10.1016/S1470-2045(04)01561-X] [PMID: 15384216]

[27]  WHO Regional Office for Europe. 2015. http://www.euro.who.int/ __data/ assets/ pdf_file/0004/276772/Economic-costhealth-impact-air-pollution-en.pdf?ua=1

[28]  Comba P, Bianchi F, Fazzo L, *et al.* Cancer mortality in an area of Campania (Italy) characterized by multiple toxic dumping sites. Ann N Y Acad Sci 2006; 1076: 449-61.
[http://dx.doi.org/10.1196/annals.1371.067] [PMID: 17119224]

[29]  Fazzo L, Belli S, Minichilli F, *et al.* Cluster analysis of mortality and malformations in the Provinces of Naples and Caserta (Campania Region). Ann Ist Super Sanita 2008; 44(1): 99-111.
[PMID: 18469382]

[30]  Fazzo L, De Santis M, Mitis F, *et al.* Ecological studies of cancer incidence in an area interested by dumping waste sites in Campania (Italy). Ann Ist Super Sanita 2011; 47(2): 181-91.
[PMID: 21709388]

[31]  De Felice B, Nappi C, Zizolfi B, *et al.* Telomere shortening in women resident close to waste landfill sites. Gene 2012; 500(1): 101-6.
[http://dx.doi.org/10.1016/j.gene.2012.03.040] [PMID: 22465532]

[32]  Giovannini A, Rivezzi G, Carideo P, *et al.* Dioxins levels in breast milk of women living in Caserta and Naples: assessment of environmental risk factors. Chemosphere 2014; 94: 76-84.
[http://dx.doi.org/10.1016/j.chemosphere.2013.09.017] [PMID: 24120012]

[33]  Rivezzi G, Piscitelli P, Scortichini G, *et al.* A general model of dioxin contamination in breast milk: Results from a study on 94 women from the Caserta and Naples areas in Italy. Int J Environ Res Public Health. 2013; 8;10(11):5953-70.

[34]  Pirastu R, Pasetto R, Zona A, *et al.* The health profile of populations living in contaminated sites: SENTIERI approach. J Environ Public Health 2013; 2013: 939267.
[http://dx.doi.org/10.1155/2013/939267] [PMID: 23853611]

[35]  Triassi M, Alfano R, Illario M, Nardone A, Caporale O, Montuori P. Environmental pollution from illegal waste disposal and health effects: A review on the "Triangle of Death". Int J Environ Res Public Health. 2015; 22;12(2):1216-36.

[36]  Fusco M. Relazione sulle attività del Registro Tumori della ASL Napoli 3 sud in merito all'Area territoriale della Regione Campana denominata "Terra dei Fuochi". Registro Tumori ASL Napoli 3 sud. 2017.

**CHAPTER 9**

# The Outbreak of the Pandemic of Coronavirus Disease 2019 and its Impact on Medical Waste Management

**Gabriella Marfe[1,*], Carla Di Stefano[2] and Giulio Tarro[3]**

[1] *Department of Scienze e Tecnologie Ambientali, Biologiche e Farmaceutiche, University of Campania "Luigi Vanvitelli," via Vivaldi 43, Caserta, Italy*

[2] *Department of Hematology, "Tor Vergata" University, Viale Oxford, Rome, Italy*

[3] *University Thomas More U.P.T.M, Rome, Italy*

**Abstract:** Today, the global outbreak of the pandemic of coronavirus disease 2019 has the potential to wreak a serious impact on human health and to further trigger a global crisis. A great repercussion of this pandemic is occurring on sustainable medical waste management practices with a profound impact. On the one hand, medical waste management companies need to be ready to assist countries worldwide as they seek to manage the great volumes of infectious material; on the other hand, the use of gloves and face masks is increasing among population to stop the spread of the novel coronavirus. In addition to potentially being a biohazard, face masks and gloves can increase the plastic pollution if the disposal directions are not proper. Therefore, the current pandemic of the novel coronavirus poses new challenges regarding the management of medical waste practices above all for health measures for employees, proper waste treatment requirements.

The aim of this review is to collect data on the different systems and solutions implemented worldwide to manage municipal waste in the current situation.

**Keywords:** COVID-19, Face mask, Gloves, Healthcare waste, Pandemic.

## THE CURRENT HISTORY OF THE COVID-19 CORONAVIRUS

Until less than 20 years ago, coronaviruses represented a viral family that caused 10 to 30% of colds during the winter. In 2002, the SARS (Severe Acute Respiratory Syndrome) coronavirus (SARS-CoV) affected 8 thousand individuals in China, causing about 10% of mortality. SARS-CoV-like viruses were detected

* **Corresponding author Gabriella Marfe:** Department of Scienze e Tecnologie Ambientali, Biologiche e Farmaceutiche, University of Campania "Luigi Vanvitelli," *via* Vivaldi 43, Caserta 81100, Italy. Tel: +39 0823 275104; Fax: +39 0823 274813; E-mails: gabriellamarfe@gmail.com

**Gabriella Marfe & Carla Di Stefano (Eds.)**

in palm civets and a raccoon dog from wild-animal markets in the Guangdong Province of China, and for this reason, it is possible that these animals could transmit the infection to human being. In 2012, Middle East respiratory syndrome (MERS), a viral respiratory disease caused by a novel coronavirus (Middle East respiratory syndrome coronavirus, or MERSCoV) that was first identified in Saudi Arabia and caused the outbreak I in 2015. In December 2019, a new coronavirus appeared in Wuhan (a province of Hubei), called COVID-19 (Coronavirus Disease 2019), which after passing into exotic animals, infected humans with marked virulence leading to the spread of a mysterious pneumonia [1]. The fundamental rules during the epidemics are those to isolate patients and to carry out the quarantine such as during the typhoid fever of Athens of 430 bc, the plague of Manzoni memory, 1600 ac and the Spanish flu of 1918, that caused more victims than the just-ended First World War. Although China has seen the SARS, the country was not able to predict that the epidemic would be morphed into a "major public health event. In this regard, the lockdown in Wuhan was declared with a delay of almost a month compared to the first cases, as well as the communication to the WHO (World Health Organization) happened after the first communication of Public Health Emergency of International Concern (PHEIC).

Furthermore, samples from the throat and nasal swabs derived from possibly infected people were sent to public health agencies to detect the virus with rapid tests. Naturally, it is important to identify the proteins produced by the viral RNA for the synthesis of specific antigens and antiviral drugs. In this context, the vaccine development will take time for safe treatment in human being. In Italy, specifically, Lombardy and Veneto regions have been affected by the coronavirus epidemic from the end of January 2020. Here, it was difficult to stop the disease because the transmission of the virus occurred, such as that of the flu virus. Furthermore, many people had symptoms, but they were not aware of being exposed to the virus. The coronavirus respiratory syndrome was moderate for the majority of cases, but many cases had reported "mild" pneumonia, according to the Chinese Center for Disease Control and Prevention statement made at the end of February after the observation of about 90 thousand cases. Moreover, 14% of confirmed cases had serious pneumonia and dyspnoea, while 5% of the patients had a pulmonary collapse, a septic shock and a deficiency of several organs (2.3% of the total). Specifically, several articles reported numerous symptoms, including fever, cough, and general malaise and occasional diarrhea in the majority of cases [2]. Patients with severe COVID-19 develop acute respiratory distress syndrome and acute lung injury, leading to morbidity and mortality caused by damage to the alveolar lumen with inflammation and pneumonia [3 - 5]. A retrospective study of 522 patients showed that 82.1% of COVID-19 cases displayed low circulating lymphocyte counts. The authors found that 499 cases had low lymphocyte count recorded. In particular, the counts of total T cells, CD4+, and CD8+ were significantly lower in the intensive care unit (ICU) patients than Non-ICU patients. Such result could suggest that there was a profound T cell loss in COVID-19 disease [6]. Furthermore, the authors demonstrated that patients over 60 years old have very low T cell numbers, indicating that TNF-α might be directly involved in inducing T cell loss in these patients. Furthermore, the authors observed that the CD8 T and CD4 T cells have higher levels of PD-1 (exhaustion markers) in ICU patients when compared with non-ICU patients [6]. Generally, the coronaviruses cause symptoms, mostly colds during the winter [7]. Today, people infected by this new COVID-19 developed severe forms, above all, if they had previous diseases or in the elderly. Less than 1% of healthy subjects died of this new SARS, while cardio patients were 10.5%, diabetics 7.3% and patients with chronic respiratory diseases, hypertension or cancer 6%. The different coronaviruses strains vary in clinical severity when they infect the human species: the classic cold coronaviruses are 229E, NL63, OC43 and HKU1. The three coronaviruses that have proven dangerous for human life were MERS-CoV, SARS-CoV and the last SARS-CoV2 which was able to kill more individuals than the other two associated with the speed of the pathological transmission for a greater number of people. One possible reason for this event is that this virus is able to link with the Angiotensin-converting enzyme 2 (ACE receptor) that is present on the ciliated epithelial cells of the upper and lower respiratory tract and in alveolar type II pneumocytes. In this way, the COVID-19 into these cells, like the primitive SARS coronavirus (2002-03) with a high affinity for the ACE receptor [7]. Although different studies showed a 76% identity in the amino acid sequence between SARS-CoV and Sars-CoV2 [8], the affinity of Sars-Cov-2 towards this receptor is 10–20- fold higher than that of SARS- CoV2 In this regard, the binding between the spike protein and ACE2 occurs through proteolytic cleavage of ACE2 by transmembrane serine protease 2 (TMPRSS2). After this linking, the enzyme converts angiotensin I into angiotensin1-9, which in turn is converted to angiotensin1-7 that acts on the Mas receptor (a G protein-coupled receptor). Specifically, this receptor is expressed in a variety of cell lineages in many tissues relevant to cardiovascular disease (including type 2 alveolar epithelial cells), to modestly lower blood pressure through vasodilation and by promoting kidney sodium and water excretion but also to attenuate inflammation through the production of nitric oxide 3. Many authors supposed that the viral entry through the binding of the SARS- CoV-2 spike protein to ACE2 could suppress ACE2 expression. This process could lead to elevated internalization and shedding of ACE2 from the cell surface, that in turn could increase levels of angiotensin II (Ang II). Such protein binds its receptor AT1 (AT1R), causing an inflammatory response in the lungs and potentially triggering direct parenchymal injury. Furthermore, ACE2 receptors are expressed in the heart (endothelium of coronary arteries, myocytes fibroblasts, epicardial adipocytes), vessels (vascular endothelial and smooth cells), gut (intestinal epithelial cells), lung (tracheal and bronchial epithelial cells, type 2 pneumocytes, macrophages), kidney (luminal surface of tubular epithelial cells), testis, brain [9 - 13]. Another study has identified a fragment of the receptor-binding domain (RBD) in SARS-CoV-2 S protein that is the most important determinant of the SARS-CoV host range [14]. In another article, Coutard *et al.* observed a peculiar furin-like cleavage sequence site (PRRARS|V) in the spike protein of the 2019-nCoV, that was also present in the MERS-CoV [15]. In a recent article, the authors have identified a D614G mutation in spike protein. Furthermore, they have compared this protein containing different aa at residue 614 (aspartic acid (SD614) and glycine SG614). They found that pseudotyped retrovirus with the mutation SG614 was more efficient to infect the ACE2-expressing cells than those with SD614 [16]. Other research is designing various diagnostic tests, antiviral agents, and vaccines on the basis of our understanding of the structure and function of the various viral proteins involved in the life cycle of this virus.

## VIRAL INTERFERENCE

To explain the situation on COVID-19 in Italy, we must consider multiple factors that may have interacted among them [8]. First of all, in the North Italy, the spread of this virus was higher than in the Center and the South Italy. Secondly, the environmental and climatological situation between the North and the South of Italy had played a crucial role in the spread of the virus as shown in different studies. For example, on March 16th 2020, an Italian group wrote a position paper in which the authors proposed a research hypothesis regarding the association between higher mortality rates due to COVID-19 and high particulate matter concentrations observed in the Northern Italy. In this regards, the authors collected daily PM10 levels from all Italian Provinces between February 9th and February 29th, considering (the average time estimated in 17 days) the spread of the COVID-19 in Italy from February 24th to March 13th (the date when the lockdown has been imposed over Italy). Furthermore, they also collected population data and daily travelling information per each province. PM10 daily limit value excess seemed to be correlated with a high infection rate in univariate analyses. In this regard, a median of 0.03 infection cases over 1000 residents occurred in less polluted provinces, while this median (0.26 cases over 1000 residents) increased in most polluted provinces. These data suggested a strong association among high concentrations of PM10, the chronic exposure and the anomalous variability of SARS-CoV-2 in Italy [17].

## THERAPY FOR CORONAVIRUS DISEASE

The story of autopsies was initially important for Italian cases. In fact, it showed that mortality did not occur for interstitial pneumonia, but above all, for an embolic thrombus mechanism of the small vessels of different vital organs and, therefore, the obvious importance for an emergency room or beds in intensive care units to use the heparin and cortisone. The intubation and ventilation of many patients with COVID-19 have played a crucial role in facing a progressive lack of oxygen debt despite high-flow oxygen therapy and bilevel positive airway pressure ventilation. Lung-protective ventilation, prone position ventilation, adequate sedation and analgesia have been crucial factors in ventilation management, as shown in Chinese experience. Furthermore, in Italy, oxygen ozone therapy was used to treat patients suffering from COVID-19. The therapies involve an association of anti-viral drugs and supportive care. Significant evidence suggests that SARS-CoV-2 infection is correlated with a pro-inflammatory status characterized by high levels of different cytokines, including interleukin (IL)1β, IL1Rα, IL-2, IL10, fibroblast growth factor (FGF), granulocyte-macrophage colony stimulating factor (GM-CSF), granulocyte-colony stimulating factor (G-CSF), interferon-γ-inducible protein (IP10), monocyte chemoattractant protein (MCP1), macrophage inflammatory protein 1 alpha (MIP1A), platelet derived growth factor (PDGF), tumor necrosis factor (TNFα) and vascular endothelial growth factor (VEGF). Other drugs were tested against SARS-CoV-2 such as Remdevisir (used for Ebola), Chloroquine (Plaquenil already known as an antimalarial drug), Fapilavir (Avigan developed by Fujifilm Toyama Chemical in 2014 in Japan for the treatment of avian influenza or novel influenza resistant to neuraminidase inhibitors) AIDS virus protease inhibitors (including such as Ritonavir and Lopinavir) and Oseltamivir (Tamiflu approved for treatment of influenza A and B).

Remdesivir is a nucleotide analog originally developed by Gilead as a drug against Ebola virus, and subsequently demonstrating its efficacy in inhibiting coronaviruses such as SARS-CoV and MERS-CoV *in vitro*. Successful case studies describing the use of Remdesivir for COVID-19 have been reported [18, 19]. A preliminary report on a randomized, placebo-controlled trial in COVID 19 patients was introduced. In this study, Beigel *et al.* showed a reduction in time to recovery from a median of 15 days among placebo recipients to 11 days among those receiving Remdesivir. Furthermore, the authors observed lower mortality among patients who received remdesivir (7.1%) than among those who received placebo (11.9%), but the difference was not statistically significant [20]. In another article. Goldman *et al.* investigated effects of Remdesivir on Covid-19 patients after 5 days and 10 days from treatment. They found the same effects of 5 days and 10 days of Remdesivir therapy, after adjustments for baseline clinical status. However, the absence of a control group in this study did not permit an overall assessment of the efficacy of such drug [21]. Clinical trials are ongoing to evaluate the safety and antiviral activity of Remdesivir in patients with mild to moderate or severeCOVID-19 (NCT04292899, NCT04292730, NCT04257656, NCT04252664, NCT04280705). In addition, the National Institutes of Health is sponsoring a new randomized, double-blind, placebo-controlled trial that will shed light on the effectiveness of Remdesivir compared with supportive care (NCT04280705) [22].

Favipiravir, previously known as T-705, is a prodrug of a purine nucleotide, Favipiravir ribofuranosyl-5'-triphosphate. The active agent are able to stop the RNA polymerase and viral replication. Most of Favipiravir's preclinical results are derived from its influenza and Ebola activity; additionally, it also inhibits other RNA viruses. The EC50 of favipiravir against SARSCoV 2 was 61.88 µM/L in Vero E6 cells (*in vitro*) [23]. In a prospective, randomized, multicenter study, the authors compared Favipiravir (n = 120) and Arbidol (n = 120) effects for the treatment of moderate and severe COVID-19 infections. The observed differences in clinical recovery after 7 days in patients with moderate infections were 71.4% Favipiravir and 55.9% Arbidol, P = .019, while there were no significant effects in the severe or severe and moderate (combined) arms [24]. Furthermore, such drug seemed to be effective in COVID-19. In other clinical trials conducted in Shenzhen, preliminary clinical data suggested that it was able to exert antiviral effects more potently than Lopinavir/ritonavir with no overt adverse reactions [25]. Although the clinical results of this drug are favorable, further studies are necessary using a large sample size. Lopinavir/ritonavir (LPV-r) is a combination of the human immunodeficiency virus (HIV)-specific protease inhibitor that is used as first-line therapy for HIV. In a Chinese trial, patients with severe COVID-19 were treated with the lopinavir/ritonavir dosing regimen at 400mg/100mg twice daily for up to 14 days [26]. In patients with COVID-19, the adverse effects of such drugs could increase for associated therapy or viral infection since 20% to 30% of patients showed a high value of transaminases at presentation [27].

Chloroquine and hydroxychloroquine are two similar drugs that are active against malaria parasite. Several *in vitro* studies and recent clinical trials have shown the efficacy of chloroquine in patients with COVID-19 at different levels of severity [28 - 30]. Such drugs seem to stop viral entry into cells by inhibiting glycosylation of host receptors, proteolytic processing, and endosomal acidification, at the same time, they are able to decrease cytokine production and inhibition [31, 32]. Other published studies showed that the replication is blocked by both drugs since they altered ACE2 glycosylation by stopping S-protein binding. These drugs are used against Zika virus, Ebola virus, and Chikungunya virus [25, 33]. Furthermore, other studies found that zinc could reduce the SARS-CoV-2 viral activities. In this regard, chloroquine could induce the uptake of zinc into the cytosol of the cell, which is capable of inhibiting RNA-dependent RNA polymerase and ultimately halting the replication of coronavirus in the host cell. Currently, there are several clinical trials in numerous countries of the world predicated upon a synergistic administration of Zn supplemented with CQ or HCQ against the novel SARS-CoV-2 virus [34]. In addition, in an open-label French study, the combination of HCQ and azithromycin showed promising results in patients with COVID19 [27, 35].

However, several studies [36, 37] found many serious side effects such as heart rhythm problems with chloroquine or hydroxychloroquine, particularly in combination with the antibiotic azithromycin. Other clinical trials are currently studying the effectiveness of chloroquine or hydroxychloroquine high doses to treat COVID-19 and the possible side effects [38]. Finally, several published articles on the potential usefulness of these drugs in COVID 19-infected patients have reported that data have been largely inconclusive for different reasons such as varying dosage regimens, disease severity, and lack of control groups.

Another class of additional treatments for COVID-19 is monoclonal antibodies directed against key inflammatory cytokines or other aspects of the innate immune response. Their use plays a crucial step to avoid significant organ damage in the lungs and other organs caused by an increased immune response and cytokine release, or "cytokine storm". IL-6 appears to be a key driver of this dysregulated inflammation based on early case series from China. Tocilizumab (TCZ) and Sarilumab are monoclonal antibodies against IL-6 receptors and have been employed to treat rheumatoid arthritis [39, 40]. TCZ and Sarilumab block IL-6 through both membrane-bound and soluble IL-6R [39, 41]. FDA has approved TCZ use during cytokine release syndrome that causes increased cytokine production and, consequently, rapid multiorgan damage (lungs, kidney, and heart) [42]. COVID-19 disease severity depends on the increase in different pro-inflammatory factors such as IL-6, IL-1, IL-2, IL-7, IL-10, granulocyte-colony stimulating factor, interferon-$\gamma$-inducible protein 10, monocyte chemoattractant protein 1, macrophage inflammatory protein-1 alpha, and TNF-$\alpha$ [43, 44]. A report showed that the treatment with tocilizumab (400mg) on 21 patients with COVID-19 treated was associated with clinical improvement in 91% of patients as measured by improved respiratory function [45]. Three clinical trials (ChiCTR2000030196, ChiCTR2000030442, and ChiCTR2000029765) for TCZ have been approved for COVID-19, and the National Health and Family Planning Commission of China has approved the treatment with TCZ in patients with elevated IL-6 level. Today, the first randomized, double-blind, placebo-controlled phase III study to investigate Actemra/RoActemra (tocilizumab) in adult patients hospitalised with severe COVID-19 associated pneumonia (locations in the US, Canada and Europe) has not reported the improvement in these patients with severe COVID-19 with pneumonia. In addition to COVACTA, different studies are investigating on Actemra/RoActemra to treat patients with COVID-19 associated pneumonia, including two phase III clinical trials, REMDACTA and EMPACTA, as well as the phase II MARIPOSA trial [46].

Other monoclonal antibodies or immunomodulatory agents in clinical trials in China or available for expanded access in the US include bevacizumab (anti–vascular endothelial growth factor medication; NCT04275414), fingolimod (immunomodulator approved for multiple sclerosis; NCT04280588), and eculizumab (antibody inhibiting terminal complement; NCT04288713). Bevacizumab is a monoclonal anti-vascular endothelial growth factor (VEGF) antibody. It is able to link with VEGF receptors on the surface of endothelial cells for VEGF binding. In this way, it blocks the binding VEGF to its receptors, and in turn, stops endothelial cell proliferation and neovascularization. Fanelli and Ranieri demonstrated the causal link between ARDS and increased vascular permeability and pulmonary edema, then Bevacizumab might be likely to be a promising therapy anti-ARDS [47]. The most effective cure (as demonstrated in an article by some Chinese virologists and published in the Proceeding National Academy of Science and in another article in the Medical Journal of Virology) is immunotherapy. That is to say, the use of gamma-globulins that are obtained from the blood of patients recovered from Sars-CoV2. It has been scientifically proven that 200 ml of transfused plasma in patients is enough to resolve the most serious situations within 48 hours [48]. In Mantua and Parma (Italy), where immunotherapy is practiced, there are already good results as shown in different studies [49]. In order to evaluate the efficacy of convalescent plasma therapy in COVID-19 patients, Ye *et al.* enrolled six laboratory confirmed COVID-19 patients to receive the transfusion of ABO-compatible convalescent plasma. After this treatment, the patients showed the alleviation of symptoms, changes in radiologic abnormalities and laboratory tests, without the onset of adverse effects. The authors concluded that this therapy could be very effective and specific for COVID-19 and represent a promising state-of-the-art therapy during the COVID-19 pandemic crisis [50]. Another study described the use of the convalescent plasma therapy in two patients affected by COVID-19. The first patient, a 71-years-old man diagnosed with COVID-19, started the therapy after 10 days after his hospitalization following a significant worsening of his general conditions and respiratory distress. The convalescent plasma was obtained from a male donor who had recovered from COVID-19 for 21 days. The second patient, a 67 years old woman affected by COVID-19 with acute respiratory distress, was treated with the convalescent plasma derived from a male donor who had recovered from COVID-19 for 18 days. Both patients presented ARDS and showed a favorable outcome after the use of convalescent plasma in addition to systemic corticosteroid [51]. Kai Duan and colleagues treated ten patients affected by severe COVID-19 with one dose of 200 mL of convalescent plasma (CP) derived from recently recovered donors. The plasma was transfused to the patients in addition to maximal supportive care and antiviral agents. After plasma transfusion, the level of neutralizing antibody rapidly increased and the clinical symptoms improved along with an increase of oxyhemoglobin saturation within 3 days. In this case, the authors observed that convalescent plasma therapy is a well-tolerated treatment [52]. Shen *et al.* reported preliminary data of 5 severely ill patients with coronavirus disease 2019 (COVID-19) treated with plasma derived from recovered individuals. All patients received mechanical ventilation for their severe respiratory failure; one patient needed extra-corporeal membrane oxygenation (ECMO) and 2 patients had bacterial and/or fungal pneumonia. The other four patients received convalescent plasma, while another patient with hypertension and mitral valve insufficiency received the plasma transfusion at day 10 [53]. The donor plasma contained IgG and IgM anti– SARS-CoV-19 antibodies that were able to neutralize the virus in *in vitro* cultures. Many trials have been registered between March and April 2020 and all of them are ongoing (Table **1**). In another very recent study, the authors reported the isolation of several SARS-CoV-2-neutralizing monoclonal antibodies (almost 61) from 5 infected patients hospitalized with severe disease. Among these are 19 antibodies that are able to neutralize the SARS-CoV-2 *in vitro*. Furthermore, epitope mapping studies found different antibodies directed to the receptor-binding domain (RBD) and those to the N-terminal domain (NTD), indicating that both of these regions at the top of the viral spike are immunogenic [54]. Some studies have evaluated the use of corticosteroids in patients with COVID-19. One paper compared two groups: the first group made up of 26 patients with severe COVID-19 treated with methylprednisolone (1–2 mg/kg/d for 5–7 days) was compared to the second group of 20 patients undergoing standard therapy. The first group reached faster improvement in clinical symptoms and lung lesions detected by CT imaging. However, two deaths occurred in the first group and one death in the second group [55]. Another paper reported a case study on one patient with COVID-19 treated with methylprednisolone since day 8 of the disease course. However, he developed respiratory failure and died on day 14 [56].

**Table 1. Clinical trials registered at ClinicalTrials.gov to determine the efficacy and safety of convalescent plasma for the treatment of patients affected by COVID-19.**

| N | NTC Number | Title | Status | Participants | Date of Start | Date of Completion | Country |
|---|---|---|---|---|---|---|---|
| 1 | NCT04333355 | Safety in Convalescent Plasma Transfusion to COVID-19 | Not yet recruiting | 20 | April 15, 2020 | December 20, 2020 | Mexico |

*(Table 1) cont.....*

| N | NTC Number | Title | Status | Participants | Date of Start | Date of Completion | Country |
|---|---|---|---|---|---|---|---|
| 2 | NCT04340050 | COVID-19 Convalescent Plasma | Recruiting | 10 | April 10, 2020 | December 31, 2020 | USA |
| 3 | NCT04343261 | NCT04343261 Convalescent Plasma in the Treatment of COVID 19 | Not yet recruiting | 15 | April 12, 2020 | April 2021 | USA |
| 4 | NCT04347681 | Potential Efficacy of Convalescent Plasma to Treat Severe COVID-19 and Patients at High Risk of Developing Severe COVID-19 | Not yet recruiting | 40 | April 12, 2020 | April 12, 2021 | USA |
| 5 | NCT04345991 | Efficacy of Convalescent Plasma to Treat COVID-19 Patients, a Nested Trial in the CORIMUNO-19 Cohort | Not yet recruiting | 120 | April 14, 2020 | April 1, 2021 | Saudi Arabia |
| 6 | NCT04346446 | Efficacy of Convalescent Plasma Therapy in Severely Sick COVID-19 Patients | Recruiting | 20 | April 14, 2020 | June 30, 2020 | India |
| 6 | NCT04342182 | Convalescent Plasma as Therapy for COVID-19 Severe SARS-CoV-2 Disease (CONCOVID Study) | Recruiting | 426 | April 18, 2020 | June 1, 2020 | Netherlands |
| 7 | NCT04345679 | Anti COVID-19 Convalescent Plasma Therapy | Not yet recruiting | 20 | April 14, 2020 | April 1, 2021 | Hungary |

*(Table 1) cont.....*

| N | NTC Number | Title | Status | Participants | Date of Start | Date of Completion | Country |
|---|---|---|---|---|---|---|---|
| 8 | NCT04343755 | Convalescent Plasma as Treatment for Hospitalized Subjects With COVID-19 Infection | Recruiting | 55 | April 9, 2020 | April 2020 | USA |
| 9 | NCT04327349 | Investigating Effect of Convalescent Plasma on COVID-19 Patients Outcome: A Clinical Trial | Enrolling by invitation | 30 | March 28, 2020 | September 30, 2020 | Iran |
| 10 | NCT04332380 | Convalescent Plasma for Patients with COVID-19: A Pilot Study | Not yet recruiting | 1200 | April 1, 2020 | December 31, 2020 | Colombia |
| 11 | NCT04332835 | Convalescent Plasma for Patients with COVID-19: A Randomized, Open Label Parallel, Controlled Clinical Study | Not yet recruiting | 80 | April 1, 2020 | December 31, 2020 | Colombia |
| 12 | NCT04345523 | Convalescent Plasma Therapy *vs.* SOC for the Treatment of COVID19 in Hospitalized Patients | Recruiting | 278 | April 3, 2020 | July, 2020 | Spain |
| 14 | NCT04344535 | Convalescent Plasma *vs.* Standard Plasma for COVID-19 | Enrolling by invitation | 500 | April 8, 2020 | August 31, 2021 | USA |
| 15 | NCT04344015g | COVID-19 Plasma Collection | Recruitin | 2000 | April 13, 2020 | April 12, 2021 | USA |
| 16 | NCT04292340 | Anti-SARS-CoV-2 Inactivated Convalescent Plasma in the Treatment of COVID-19 | Recruiting | 15 | February 1, 2020 | December 31, 2020 | China |

*(Table 1) cont.....*

| N | NTC Number | Title | Status | Participants | Date of Start | Date of Completion | Country |
|---|---|---|---|---|---|---|---|
| 17 | NCT04334876 | Rapid SARS-CoV-2 IgG Antibody Testing in High Risk Healthcare Workers | Not yet recruiting | 340 | April 1, 2020 | January 1, 2021 | USA |
| 18 | NCT043238001 | Efficacy and Safety Human Coronavirus Immune Plasma (HCIP) *vs.* Control (SARS-CoV-2 Non-immune Plasma) Among Adults Exposed to COVID-19 | Not yet recruiting | 150 | May 1, 2020 | January, 2023 | USA |
| 19 | NCT04345289 | Efficacy and Safety of Novel Treatment Options for Adults With COVID-19 Pneumonia | Not yet recruiting | 1500 | April 20, 2020 | June 15, 2021 | Netherland |
| 20 | NCT04346589 | Convalescent Antibodies Infusion in Critically Ill COVID 19 Patients | Not yet recruiting | 10 | April, 2020 | June, 2020 | Italy |
| 21 | NCT04348877 | Plasma Rich Antibodies from Recovered Patients from COVID19 | Not yet recruiting | 20 | April 20, 2020 | December, 2020 | Egypt |
| 22 | NCT04344977 | COVID-19 Plasma Collection | Not yet recruiting | 2800 | April 15, 2020 | April 1, 2025 | USA |
| 23 | NCT04342195 | Acquiring Convalescent Specimens for COVID-19 Antibodies | Recruiting | 12 | March 5, 2020 | March, 2021 | USA |

Furthermore, another therapy, such as the complement C3 inhibitor AMY-101 (Amyndas Pharmaceuticals, Glyfada, Greece) and the anti-C5 monoclonal antibody eculizumab (Soliris; Alexion, Boston, MA), was administered in ill COVID-19 patients. The AMY-101 is able to inhibit the cleavage of C3 that, in turn, prevents the formation of the C3 and C5 convertases and the subsequent release of the inflammatory mediators C3a and C5a and formation of the tissue-damaging membrane attack complex (MAC; C5b-9). The eculizumab is able to block the cleavage of C5 and the formation of the inflammatory anaphylatoxin C5a and of the MAC/C5b-9 [57]. In Italy, recent studies have reported that the levels of the C5 activation fragment C5a and soluble MAC (sC5b-9) were elevated in the plasma of patients with severe COVID-19 by confirming that C5 blockade could represent a potential treatment [58]. Others complement inhibitors are under consideration for compassionate use in COVID-19 such as Avdoralimab (Innate Pharma, Marseille, France, an anti-C5aR monoclonal antibody that blocks binding of C5a to its receptor C5aR, CD88), or IFX-1 (InflaRX; Martinsried, Germany a monoclonal antibody that targets C5a, preventing it from interacting with the C5aR). Moreover, the recombinant human C1 esterase inhibitor conestat alfa (Ruconest; Pharming Group & Salix Pharmaceuticals, Bridgewater, NJ specific inhibitor of the classical complement activation pathway) is under consideration for an open-label, multicenter pilot trial in adult patients with SARS-CoV-2 pneumonia "PROTECT-COVID-19" trial. In a paper, the authors found in the histopathological samples of the microvasculature of human lung tissue specimens derived from a patient with COVID-19, the deposition of complement lectin pathway components MBL and MASP-2 and also complement activation fragments C4d and C5b-9. Specifically, in this study, lung and cutaneous tissues from 5 patients with SARS-CoV-2 infection with severe respiratory failure were examined. The authors observed that three patients had a consistent systemic procoagulant state (such as retiform purpura or livedo racemosa) and dermatologic signs of a generalized microvascular thrombotic disorder with elevated d-dimers level [59]. Other authors observed in other patients that the mannose binding lectin (MBL) activated the complement cascade forming complexes with MBL-associated serine proteases 1 and 2 (MASP-1 and MASP-2, respectively) in a similar manner when C1 was triggered by the classical pathway that leads to C4 and C2 cleavage and assembly of the C3 convertase [60]. In addition, other Chinese authors reported diffuse alveolar damage (DAD) with edema, hyaline membranes, and inflammation, followed by type II pneumocyte hyperplasia, and feature characteristics of typical ARDS in COVID19 patients [61, 62]. Nevertheless, future randomized controlled trials are necessary to confirm these findings and also further study the mid- and long-term outcomes after discharge.

## THE GLOBAL MEDICAL WASTE MANAGEMENT DURING THE COVID 19 PANDEMIC

During a pandemic disease outbreak, the generation of infectious medical waste, as well as the other healthcare hazards, increases in an enormous amount within a very short period due to the exponentially rapid spread of the disease at the initial stage. For this reason, the safe and fast planning of medical waste management plays a crucial role during this short time, because normally it occurs from several weeks to several months. In this context, temporary facilities need to be installed in a timely and responsive way to collect and store the great amount of medical waste. Naturally, it is necessary to adopt a programming model for effective management of medical waste in a pandemic outbreak. Such a model should have different approaches 1) the implementation of some temporary facilities 2) the suitable transportation strategies to face a tremendous increase of medical waste and 3) the proper facilities for collection, transportation and treatment of medical waste and healthcare hazards. In addition, the risk from the residues of medical waste should be reduced at treatment centers, but also the store, transportation and disposal should be safely in later stages. Furthermore, it could be important to choose the safe methods such as the radio-frequency identification (RFID) that can track waste materials to ensure safe management and disposal, and to avoid the illegal dumping.

Medical waste in hospitals should be tracked to improve their collection and transportation. In this emergency, the use of gloves and face masks is increased among health professionals. In this regard, different types of gloves must be disposed of in a proper manner. For example, latex gloves are a biodegradable product since the latex substance is obtained from the rubber trees. At the end of their use, they are disposed of a landfill or waste-to-energy plants, but they are able to decompose within a few months. Other gloves are composed of nitrile that is a very elastic synthetic rubber with mechanical and chemical resistance. They are produced from an organic compound and, for this reason, cannot be recycled. Furthermore, polyvinyl chloride gloves (known to most with the abbreviation PVC) must be given in the separate collection of plastic. Such a compound is chemically obtained by material with the addition of plasticizers. This type of gloves would release substances dangerous for the environment and human health when they are destined to incinerators. For example, PVC, during the combustion phase, produces several chemical compounds of the dioxin family that are carcinogenic. For these reasons, vinyl gloves must be destined for the collection of plastic and they are treated in special recycling plants to obtain new products, or also heat and electricity (Fig. **1**). Not only single-use gloves but also the use of face masks is increasing a lot during this pandemic. Many people wear masks to protect themselves from the risk of infection when they go to work or go shopping at the supermarket. The face masks are not recyclable and they are disposed of the unsorted waste. In this regard, the U.S. Food and Drug Administration has approved a product of the Critical Care Decontamination System™ manufactured by Battelle Memorial Institute and in Duke Health's recycling of N95 masks using vaporized hydrogen peroxide.

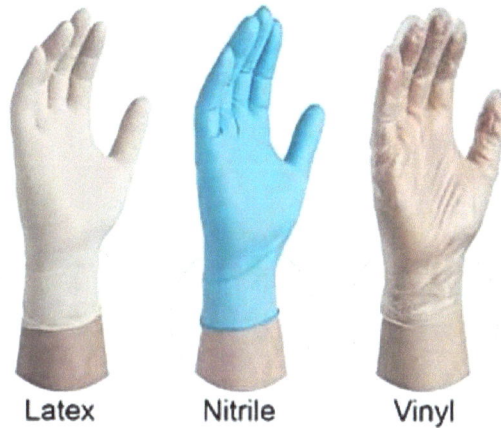

Latex      Nitrile      Vinyl

**Fig. (1).**  Types of gloves.

In Hong Kong, a local public research institute has developed reusable masks that employ enhanced filtration technology that allows their washing and reusing up to 60 times. Such masks earned top honours at a European exhibition in 2018. Certain batches have already been reserved for local schools, with senior secondary students likely to resume classes in May 2020. In Italy, Capannori has reached up to 40% of waste reduction and 82% of waste segregation [63]. Capannori is one of the highest in Europe, supported by good waste management policies and high public awareness. A disposable mask has been developed by the Lucca design agency Oroburro in collaboration with the Research Center Zero Capannori and Zero Waste Italy waste in an interesting project, called Arya. This mask is not suitable for hospitals and but is perfect for everyday life and it is made of common textile, usually cotton, cellulose based filter compound commonly used in the food industry. Today, it is in the validation phase for bacterial filtering capacity. This is a good solution that allows the whole population to be equipped to contain the spread of the virus and, at the same time, prevent an environmental disaster . (https://www.toscanachiantiambiente. it/ capannori-il-centro-rifiuti-zero -lancia-la-eco-mascherina-che-non-inquina/).

In this context, an effective and efficient program for the management of healthcare waste is an important component in the COVID-19 hospital.

The starting point of implementation of the biomedical waste management program (Fig. **2**) within the COVID 19 hospital should follow the most important legislative in the field of biomedical waste. In this regard, the framework program of biomedical waste management should include the identification of the waste categories and the strategy to reduce the waste amount, such as:

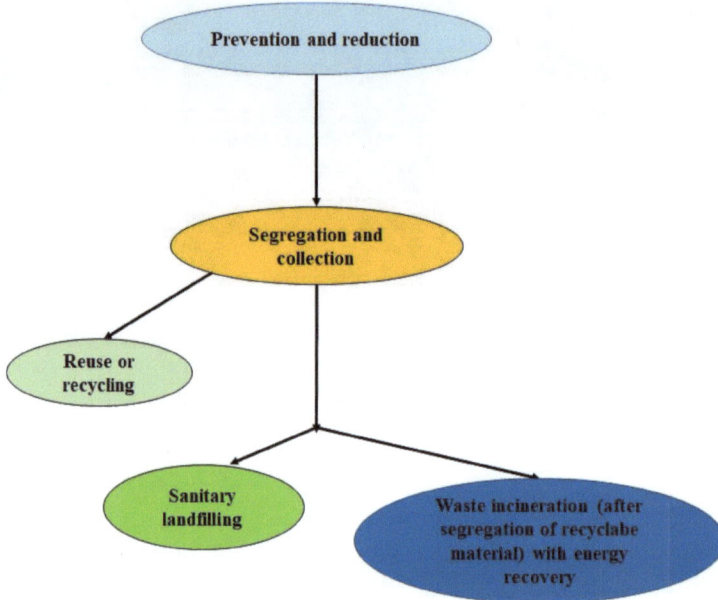

**Fig. (2).** Biomedical waste management.

a. Input data:
Documentation of equipment and medication
  - Information on diagnostic and analysis equipment
  - Legislation and regulations
b. Activities in the process:
  ○ Environmental aspects and impacts
  ○ Environmental impact assessment
  ○ Environmental aspects database
    The environmental analysis must be done:

Initially, the biomedical safe waste management system must be decided. Furthermore, changes in the organization (including new departments, necessary materials, equipment, number of tests made, new regulations in the environmental and sanitation field) may influence the initial analysis. Moreover, the Clinic Head of Department should consider input-output analysis to face every problem. For instance, medicines, equipment, hazardous substances should enter as inputs materials, while expired medicines, laboratory calibration equipment should be included in outputs materials. Furthermore, the Environmental officer should be received this information to submit every detail to the Environmental Manager for the approval of the Hospital Director. Specifically, it should be necessary to describe all operating conditions such as normal (N), abnormal (A) and urgency (U). In this regard, the environmental officer should be able to determine the environmental impact of medical waste on air, water, soil. Decision-makers must be able to manage medical waste in their COVID-19 strategies. Each country's ability to face this waste management can depend on the combination of different factors, such as policymaking and enforcement, existing collection, transportation, management facilities and dominant medical waste treatment. In this context, the policy will play a critical role to face and overcome this crisis

In this scenario, every country must be able to handle such a large global influx of waste by increasing the facility capacities through new technologies (such as tracking method) to ensure sustainable medical waste management. In this context, the public and private medical waste management companies should be able to face this new challenge. It will be important to manage supply chains and distribution channels. Furthermore, it will be necessary to improve waste infrastructures and enforcing sanitation practices to maximize the global market impact.

## Our Reflection

We wish to know the infection rate of COVID-19 among people living close to hazardous dumpsites. In fact, chemicals derived from dumps can damage immune systems, especially in children, and cause serious illnesses, including cancers. We think that people with weakened immune systems and other health issues are more at risk of contracting coronavirus.

*Gabriella Marfe and Carla Di Stefano*

## CONSENT FOR PUBLICATION

Not applicable.

## CONFLICT OF INTEREST

There are no conflict of interest.

## ACKNOWLEDGEMENTS

Declared none.

## REFERENCES

[1]　Tarro G. The new coronavirus from the chinese city of Wuhan. Int J Recent Sci Res 2020; 11(1): 36901-2.

[2]　Chen N, Zhou M, Dong X, *et al.* Epidemiological and clinical characteristics of 99 cases of 2019 novel

coronavirus pneumonia in Wuhan, China: a descriptive study. Lancet 2020; 395(10223): 507-13.
[http://dx.doi.org/10.1016/S0140-6736(20)30211-7] [PMID: 32007143]

[3]     Wölfel R, Corman VM, Guggemos W, *et al.* Virological assessment of hospitalized patients with COVID-2019. Nature 2020; 581(7809): 465-9.
[http://dx.doi.org/10.1038/s41586-020-2196-x] [PMID: 32235945]

[4]     Xu Z, Shi L, Wang Y, *et al.* Pathological findings of COVID-19 associated with acute respiratory distress syndrome. Lancet Respir Med 2020; 8(4): 420-2.
[http://dx.doi.org/10.1016/S2213-2600(20)30076-X] [PMID: 32085846]

[5]     Wang D, Hu B, Hu C, *et al.* Clinical characteristics of 138 hospitalized patients with 2019 novel coronavirus-infected pneumonia in Wuhan, China. JAMA. 2020 7; 323(11): 1061-9.

[6]     Diao B, Wang C, Tan Y, *et al.* Reduction and Functional Exhaustion of T Cells in Patients With Coronavirus Disease 2019 (COVID-19). Front Immunol 2020; 11: 827.
[http://dx.doi.org/10.3389/fimmu.2020.00827] [PMID: 32425950]

[7]     Tarro G. Pathogenesis of COVID-19 and the body's responses. Int J Recent Sci Res 2020; 11: i37940-2.

[8]     Tarro G. Further news on coronavirus COVID-19 from Italy. Int J Curr Res 2020; 12: 10260-3.

[9]     Kuba K, Imai Y, Penninger JM. Multiple functions of angiotensin-converting enzyme 2 and its relevance in cardiovascular diseases. Circ J 2013; 77(2): 301-8.
[http://dx.doi.org/10.1253/circj.CJ-12-1544] [PMID: 23328447]

[10]    Fukui K, Yang Q, Cao Y, *et al.* The HNF-1 target collectrin controls insulin exocytosis by SNARE complex formation. Cell Metab 2005; 2(6): 373-84.
[http://dx.doi.org/10.1016/j.cmet.2005.11.003] [PMID: 16330323]

[11]    Gallagher PE, Ferrario CM, Tallant EA. Regulation of ACE2 in cardiac myocytes and fibroblasts. Am J Physiol Heart Circ Physiol 2008; 295(6): H2373-9.
[http://dx.doi.org/10.1152/ajpheart.00426.2008] [PMID: 18849338]

[12]    Warner FJ, Lew RA, Smith AI, Lambert DW, Hooper NM, Turner AJ. Angiotensin-converting enzyme 2 (ACE2), but not ACE, is preferentially localized to the apical surface of polarized kidney cells. J Biol Chem 2005; 280(47): 39353-62.
[http://dx.doi.org/10.1074/jbc.M508914200] [PMID: 16166094]

[13]    Tseng CT, Tseng J, Perrone L, Worthy M, Popov V, Peters CJ. Apical entry and release of severe acute respiratory syndrome-associated coronavirus in polarized Calu-3 lung epithelial cells. J Virol 2005; 79(15): 9470-9.
[http://dx.doi.org/10.1128/JVI.79.15.9470-9479.2005] [PMID: 16014910]

[14]    Tai W, He L, Zhang X, *et al.* Characterization of the receptor-binding domain (RBD) of 2019 novel coronavirus: implication for development of RBD protein as a viral attachment inhibitor and vaccine. Cell Mol Immunol 2020; 17(6): 613-20.
[http://dx.doi.org/10.1038/s41423-020-0400-4] [PMID: 32203189]

[15]    Coutard B, Valle C, de Lamballerie X, Canard B, Seidah NG, Decroly E. The spike glycoprotein of the new coronavirus 2019-nCoV contains a furin-like cleavage site absent in CoV of the same clade. Antiviral Res 2020; 176: 104742.
[http://dx.doi.org/10.1016/j.antiviral.2020.104742] [PMID: 32057769]

[16]    Daniloski Z, Guo X, Sanjana NE. The D614G Mutation in SARS-CoV-2 Spike Increases Transduction of Multiple Human Cell Types. Version 2. bioRxiv. 2020 Jun 15:2020.06.14.151357. Preprint. https://www.ncbi.nlm.nih.gov/pmc/articles/PMC7310625/ Preprint.
[http://dx.doi.org/10.1101/2020.06.14.151357] [PMID: 32587969]

[17]    Setti L Passarini F, De Gennaro G, Barbieri P, *et al.* Potential role of particulate matter in the spreading of COVID-19 in Northern Italy: First observational study based on initial epidemic diffusion. BMJ Open 2020; 10(9): e039338.

[http://dx.doi.org/10.1136/bmjopen-2020-039338]

[18]  Holshue ML, DeBolt C, Lindquist S, *et al.* Washington State 2019-nCoV Case Investigation Team. First Case of 2019 Novel Coronavirus in the United States. N Engl J Med 2020; 382(10): 929-36. [http://dx.doi.org/10.1056/NEJMoa2001191] [PMID: 32004427]

[19]  Kujawski SA, Wong K, Collins JP, *et al.* COVID-19 Investigation Team. Clinical and virologic characteristics of the first 12 patients with coronavirus disease 2019 (COVID-19) in the United States. Nat Med 2020; 26(6): 861-8. [http://dx.doi.org/10.1038/s41591-020-0877-5] [PMID: 32327757]

[20]  Beigel JH, Tomashek KM, Dodd LE, *et al.* ACTT-1 Study Group Members. Remdesivir for the Treatment of Covid-19 — preliminary report. N Engl J Med 2020. 22: NEJMoa2007764. [http://dx.doi.org/10.1056/NEJMoa2007764] [PMID: 32445440]

[21]  Goldman JD, Lye DCB, Hui DS, *et al.* GS-US-540-5773 Investigators. Remdesivir for 5 or 10 days in patients with severe Covid-19. N Engl J Med 2020. May 27;NEJMoa2015301. [http://dx.doi.org/10.1056/NEJMoa2015301] [PMID: 32459919]

[22]  FDA issuance of emergency use authorization for potentialCOVID-19 treatment. Press release of the Food and Drug Administration, May 1, 2020.

[23]  Wang M, Cao R, Zhang L, Yang X, Liu J, Xu M. Remdesivir and chloroquine effectively inhibit the recently emerged novel coronavirus (2019-nCoV) *in vitro*. Cell Res2 020 2019; 30(3): 269-7. 1

[24]  Chen C, Huang J, Cheng Z, *et al.* Favipiravir *versus* Arbidol for COVID-19: a randomized clinical trial. Trial Med Rxiv 2020 . [http://dx.doi.org/10.1101/2020.03.17.20037432]

[25]  Third People's Hospital of Shenzhen. February 14, 2020. Familavir is more effective than Lopinavir/ritonavir in the treatment of COVID-19 (in Chinese). 2020. http://www.szd syy.com/News/0a6c1e58-e3 d0-4cd1-867a-d5524bc59cd6.html

[26]  Cao B, Wang Y, Wen D, *et al.* A trial of lopinavir-ritonavir in adults hospitalized with severe COVID-19. N Engl J Med 2020; 382(19): 1787-99. [http://dx.doi.org/10.1056/NEJMoa2001282] [PMID: 32187464]

[27]  Wu C, Chen X, Cai Y, *et al.* Risk Factors Associated With Acute Respiratory Distress Syndrome and Death in Patients With Coronavirus Disease 2019 Pneumonia in Wuhan, China. JAMA Intern Med 2020; 180(7): 934-43. Epub ahead of print [http://dx.doi.org/10.1001/jamainternmed.2020.0994] [PMID: 32167524]

[28]  Wang M, Cao R, Zhang L, *et al.* Remdesivir and chloroquine effectively inhibit the recently emerged novel coronavirus (2019-nCoV) *in vitro*. Cell Res 2020; 30(3): 269-71. [http://dx.doi.org/10.1038/s41422-020-0282-0] [PMID: 32020029]

[29]  Touret F, de Lamballerie X. of chloroquine and COVID-19. Antivir Res 2020; 177 Epub 2020 Mar 5. Last accessed 4 April, 2020. [http://dx.doi.org/10.1016/j. antiviral.2020.104762.]

[30]  Liu Xi, Chen Huili, Shang Yuqi, *et al.* Efficacy of Chloroquine *versus* Lopinavir/Ritonavir in mild/general COVID-19: a prospective, open-label, multicenter randomized controlled clinical study. Trials 2020; 21(1): 622. [http://dx.doi.org/10.1186/s13063-020-04478-w.10.]

[31]  Zhou D, Dai SM, Tong Q. COVID-19: a recommendation to examine the effect of hydroxychloroquine in preventing infection and progression. J Antimicrob Chemother 2020; 75(7): 1667-70. [published online March 20, 2020]. [http://dx.doi.org/10.1093/jac/dkaa114] [PMID: 32196083]

[32]  Devaux CA, Rolain JM, Colson P, Raoult D. New insights on the antiviral effects of chloroquine against coronavirus: what to expect for COVID-19? Int J Antimicrob Agents 2020; 55(5)105938 [http://dx.doi.org/10.1016/j.ijantimicag.2020.105938] [PMID: 32171740]

[33] Li C, Zhu X, Ji X, *et al.* Chloroquine, a FDA-approved drug, prevents Zika virus infection and its associated congenital microcephaly in mice. EBioMedicine 2017; 24: 189-94.
[http://dx.doi.org/10.1016/j.ebiom.2017.09.034] [PMID: 29033372]

[34] Shittu MO, Afolami OI. Improving the efficacy of Chloroquine and Hydroxychloroquine against SARS-CoV-2 may require Zinc additives. A better synergy for future COVID-19 clinical trials. Infez Med 2020; 28(2): 192-7.

[35] Gautret P, Lagier JC, Parola P, *et al.* Hydroxychloroquine and azithromycin as a treatment of COVID-19: results of an open-label non-randomized clinical trial. Int J Antimicrob Agents 2020; 56(1)105949
[http://dx.doi.org/10.1016/j.ijantimicag.2020.105949] [PMID: 32205204]

[36] Lane JCE, Weaver J, Kosta K, *et al.* OHDSI-COVID-19 consortium. Risk of hydroxychloroquine alone and in combination with azithromycin in the treatment of rheumatoid arthritis: A multinational, retrospective study. Lancet Rheumatol Lancet Rheumatol 2020.
[http://dx.doi.org/10.1016/S2665-9913(20)30276-9] [PMID: 32864627]

[37] Tang W, Cao Z, Han M, *et al.* Hydroxychloroquine in patients with COVID-19: an open-label, randomized, controlled trial. BMJ 2020; 369: m1849.
[http://dx.doi.org/10.1136/bmj.m1849] [PMID: 32409561]

[38] Chen Z, Hu J, Zhang Z, *et al.* Efficacy of hydroxychloroquine in patients with COVID-19: results of a randomized clinical trial. Epidemiology 2020. Available from:
[http://dx.doi.org/10.1101/2020.03.22.20040758]

[39] Raimondo MG, Biggioggero M, Crotti C, Becciolini A, Favalli EG. Profile of sarilumab and its potential in the treatment of rheumatoid arthritis. Drug Des Devel Ther 2017; 1593-603.
[http://dx.doi.org/10.2147/DDDT.S100302] [PMID: 28579757]

[40] Nishimoto N, Sasai M, Shima Y, *et al.* Improvement in Castleman's disease by humanized anti-interleukin-6 receptor antibody therapy. Blood 2000; 95(1): 56-61.
[http://dx.doi.org/10.1182/blood.V95.1.56] [PMID: 10607684]

[41] Sakkas LI. Spotlight on tocilizumab and its potential in the treatment of systemic sclerosis. 2016; 10: 2723-8.
[http://dx.doi.org/10.2147/DDDT.S99696] [PMID: 27621593]

[42] Shimabukuro-Vornhagen A, Gödel P, Subklewe M, *et al.* Cytokine release syndrome. J Immunother Cancer 2020; 6(1): 56.
[http://dx.doi.org/10.1186/s40425-018-0343]

[44] Ruan Q, Yang K, Wang W, Jiang L, Song J. Clinical predictors of mortality due to COVID-19 based on an analysis of data of 150 patients from Wuhan, China. Intensive Care Med 2020; 46(5): 846-8.
[http://dx.doi.org/10.1007/s00134-020-05991-x] [PMID: 32125452]

[45] Xu X, Han M, Li T, *et al.* Effective treatment of severe COVID-19 patients with tocilizumab. Proc Natl Acad Sci USA 2020; 117(20): 10970-5.
[http://dx.doi.org/10.1073/pnas.2005615117] [PMID: 32350134]

[46] Roche provides an update on the phase III COVACTA trial of Actemra/RoActemra in hospitalised patients with severe COVID-19 associated pneumonia. https://www.roche.com/investors/updates/inv-update-2020-07-29.htm

[47] Fanelli V, Ranieri VM. Mechanisms and clinical consequences of acute lung injury. Ann Am Thorac Soc 2015; 12 (Suppl. 1): S3-8.
[http://dx.doi.org/10.1513/AnnalsATS.201407-340MG] [PMID: 25830831]

[48] Chen L, Xiong J, Bao L, Shi Y. Convalescent plasma as a potential therapy for COVID-19. Lancet Infect Dis 2020; 20(4): 398-400.
[http://dx.doi.org/10.1016/S1473-3099(20)30141-9] [PMID: 32113510]

[49] Franchini M, Marano G, Velati C, Pati I, Pupella S, Maria Liumbruno G. Operational protocol for

donation of anti-COVID-19 convalescent plasma in Italy. Vox Sang 2020. [published online ahead of print, 2020 Apr 23].
[http://dx.doi.org/10.1111/vox.12940] [PMID: 32324899]

[50]   Ye M, Fu D, Ren Y, *et al.* Treatment with convalescent plasma for COVID-19 patients in Wuhan, China. J Med Virol 2020. [Online ahead of print.].
[http://dx.doi.org/10.1002/jmv.25882] [PMID: 32293713]

[51]   Ahn JY, Sohn Y, Lee SH, *et al.* Use of Convalescent Plasma Therapy in Two COVID-19 Patients with Acute Respiratory Distress Syndrome in Korea. J Korean Med Sci 2020; 35(14)e149
[http://dx.doi.org/10.3346/jkms.2020.35.e149] [PMID: 32281317]

[52]   Duan K, Liu B, Li C, *et al.* Effectiveness of convalescent plasma therapy in severe COVID-19 patients. Proc Natl Acad Sci USA 2020; 117(17): 9490-6.
[http://dx.doi.org/10.1073/pnas.2004168117] [PMID: 32253318]

[53]   Shen C, Wang Z, Zhao F, *et al.* Treatment of 5 Critically Ill Patients With COVID-19 With Convalescent Plasma. JAMA 2020; 323(16): 1582-9. [Online ahead of print.].
[http://dx.doi.org/10.1001/jama.2020.4783] [PMID: 32219428]

[54]   Liu L, Wang P, Nair MS, *et al.* Potent neutralizing antibodies directed to multiple epitopes on SARS-CoV-2 spike. Nature 2020; 584: 450-6.
[http://dx.doi.org/10.1038/s41586-020-2571-7] [PMID: 32698192]

[55]   Wang Y, Jiang W, He Q, *et al.* A retrospective cohort study of methylprednisolone therapy in severe patients with COVID-19 pneumonia Signal Transduct Target Ther. 2020; 5(1): 57.

[56]   Polak SB, Van Gool IC, Cohen D, von der Thüsen JH, van Paassen J. A systematic review of pathological findings in COVID-19: a pathophysiological timeline and possible mechanisms of disease progression. Mod Pathol 2020; 22: 1-11.
[http://dx.doi.org/10.1038/s41379-020-0603-3] [PMID: 32572155]

[57]   Risitano AM, Mastellos DC, Huber-Lang M, *et al.* Complement as a target in COVID-19? Nat Rev Immunol 2020; 20(6): 343-4.
[http://dx.doi.org/10.1038/s41577-020-0320-7] [PMID: 32327719]

[58]   Cugno M, Meroni PL, Gualtierotti R, *et al.* Complement activation in patients with COVID-19: A novel therapeutic target. J Allergy Clin Immunol 2020; 146(1): 215-7.
[http://dx.doi.org/10.1016/j.jaci.2020.05.006] [PMID: 32417135]

[59]   Magro C, Mulvey JJ, Berlin D, *et al.* Complement associated microvascular injury and thrombosis in the pathogenesis of severe COVID-19 infection: A report of five cases. Transl Res 2020; 220: 1-13.
[http://dx.doi.org/10.1016/j.trsl.2020.04.007] [PMID: 32299776]

[60]   LaFon DC, Thiel S, Kim YI, Dransfield MT, Nahm MH. Classical and lectin complement pathways and markers of inflammation for investigation of susceptibility to infections among healthy older adults. Immun Ageing 2020. eCollection 2020.
[http://dx.doi.org/10.1186/s12979-020-00189-7]

[61]   Xu Z, Shi L, Wang Y, *et al.* Pathological findings of COVID-19 associated with acute respiratory distress syndrome. Lancet Respir Med 2020; 8(4): 420-2.
[http://dx.doi.org/10.1016/S2213-2600(20)30076-X] [PMID: 32085846]

[62]   Zhang H, Zhou P, Wei Y, *et al.* Histopathologic changes and SARS-Cov-2 immunostaining in the lung of a patient with COVID-19. Ann Intern Med 2020; 172(9): 629-32.
[http://dx.doi.org/10.7326/M20-0533.]

[63]   an Vliet A. Zero waste europe case study 1. In: Europe ZW, Ed. The Story of Capannori. Netherlands: Zero Waste Europe 2014. https://zerowasteeurope.eu/downloads/case-study-1-the-story-of-capannori/

# SUBJECT INDEX